Electricity Vs Fire

The Fight For Our Future

By Walt Patterson

To Don & Margot with best wishes

Walt

Published by Walt Patterson
Little Rushmoor High Bois Lane
Amersham Bucks
HP66DQ UK
walt@waltpatterson.org
www.waltpatterson.org

ISBN 978-0-9932612-0-6

Cover design & artwork by Samuel Dexter Davis

Remembering my beloved Cleone -
For Perd and Tab with love and hope

Table Of Contents

Prologue: Fire Hazard

Part 1: What We Do And How We Do It

Then
Now
What we do
Survival and comfort: controlling heat flow
Heating things
Making light: with fire
Exerting force: without fire
Exerting force: with fire
Moving things: without fire
Moving things: with fire
Gas light: a fire system
Electricity: a new process
Electric telegraphy: managing information
Arc light: light without fire
Incandescent light: an electricity system
Exerting force: with electricity
Moving things: with electricity
Fire inside
Moving things: fire versus electricity
Fire at sea
Cooling things
Managing information: with electricity
Electricity expanding
Fire in the air
Electric diversity to electric monopoly
War, fire, electricity
Electricity regrouping
Nuclear 'fire'

'The Stone Age did not end because we ran out of stones.'

*- attributed to Sheikh Zaki Yamani,
former Oil Minister, Saudi Arabia*

Prologue: Fire Hazard

Have you used fire today? Since you're reading this, the answer is Yes. You may not have seen a flame. But if you started the car you used fire, burning petrol in the engine. If you turned up the gas central heating you used fire, burning natural gas in the boiler. If you turned on the light, the television or the computer, the electricity may have come from a power station using fire, burning coal, natural gas or oil. If you are reading this as an electronic document on a computer, tablet, or phone, you have charged it using electricity that was probably generated using fire. Moreover, making the materials in the car, the central heating, the appliances and the building enclosing them - the metals, the plastics, the bricks, the concrete - used fire, in smelters, refineries, chemical plants, kilns, cement plants and other factories. Most of the time, in rich countries, we don't even see the fire. But it's there, almost everywhere, in our everyday lives.

We evolved with fire. Indeed fire came before us. The Fire Age began in the early Stone Age, long before *Homo sapiens*. Our Neanderthal precursors used fire. We in turn have long taken fire for granted. Most of the time, in rich countries, we no longer even realize when we use it. But we do, every day. Directly or indirectly, all over the world, we use fire all the time. We think of fire, if we think of it at all, as a key to civilization. If you ask people what they consider the greatest human achievements, the answers will probably include the wheel and the control of fire. Even today we think of fire as cosy and welcoming, the emblem of hearth and home. For most of human evolution, fire has indeed been essential and invaluable. But its greatest contribution may now be one most people might not yet recognize. Fire has made possible the human control of electricity. In turn, electricity - the right kind of electricity - may save us from fire.

Why should that be necessary? Why do we need to be saved from fire? Think about it. Why can't you breathe in Beijing? Why are governments fighting over the Arctic seabed? Why do scientists warn of ever more extreme weather worldwide? All these urgent issues have a common cause - fire. Fire in building heaters, factory furnaces, vehicle engines and power stations poisons the air in major cities around the world. The insistent need to feed fire is why governments worry and wrangle about secure supplies of fuel. Fire produces the carbon dioxide disturbing the atmosphere. We have let fire get out of control. For all its usefulness, cosiness and immediate appeal, fire is actually violent and extreme, a primitive brute-force process. It produces heat far too hot to touch. You know this from personal, perhaps uncomfortable, experience. Fire consumes the fuel that feeds it, turning it rapidly into waste. Much of this waste is unpleasant. Some is toxic. Some even upsets planetary systems. Everyone knows fire can be dangerous. That's why we tell toddlers not to play with fire, why we buy fire insurance. But the dangers from fire are now no longer merely local, of personal injury or damage to property. Fire makes air pernicious. It spews particles that clog lungs, inviting asthma and cancer. It releases gases that turn waterways acid and kill the forests around them.

Worst of all is the commonest gas that fire produces, carbon dioxide. It is not toxic, and even now it makes up only a tiny fraction of the earth's atmosphere. But in the atmosphere it has a significant effect. Sunlight warms the Earth. The Earth in turn, like any warm body, gives off invisible heat. The Earth gives off 'earthlight', sending it back into the coolness of space. The carbon dioxide in the atmosphere, however, reflects some of this earthlight, bouncing it back to the Earth. Without the heat reflected by carbon dioxide, water vapour and other gases, the Earth would not be warm enough to support life. Now, however, we use fire so much that the carbon dioxide is building up in the atmosphere, reflecting more and more heat back to the Earth. The fire we use is relentlessly heating up the planet, upsetting the atmosphere and warming the oceans. We are seeing the results all around us - violent storms, hurricanes and tornadoes, floods, heatwaves, droughts, raging wildfires, vanishing ice sheets and melting glaciers. Fire is no longer merely dangerous. The way we use fire now threatens the future of our human civilization.

That is where electricity comes in. Fire is a chemical process. It rapidly degrades what it happens in, turning it to ash and other wastes, including smoke, soot and invisible gases. Electricity, however, is not a chemical process. It is a physical process. It does not consume what it is happening in, nor does it produce dangerous waste. As a process, electricity, too, can be violent and extreme. The lightning that first gave our Neanderthal forerunners fire is nature's own electricity, raw and vivid. But we now know that electricity under human control can also be subtle and delicate, like the currents in a microchip. The process of electricity can function at any temperature, down almost to absolute zero. Electricity, moreover, is endlessly versatile. Almost everything we do with fire we can now do as well or better with electricity.

Fire, however, has a long head start. It has shaped human affairs for many millennia. Human society still relies on fire for most of its activities - even when it need not, and despite the ever more urgent problems fire is creating. We learned to control electricity only two centuries ago. Fire still dominates. Indeed we still generate most of our electricity using fire. But that is changing. The roles of fire and electricity in human activities are now evolving faster than ever before.

Can we leave the Fire Age behind? Will electricity supersede fire in time? The answer may determine our future on earth.

Part 1

What We Do and How We Do It

Then

You are a Neanderthal, although you don't know that. You, your mate, your youngsters and your tribe, 19 of you in all, live in a cave on the edge of a clearing in what will become, in 125 000 years, central Europe. You are warm-blooded, a mammal. Digesting the food you eat, mostly meat, lets your body maintain a stable internal temperature whatever the temperature around you, keeping your brain and other organs functioning smoothly. But unlike the other mammals with which you share the woods and fields, you have little or no fur outside your skin - a few patches and a light sprinkle covering only your head and not much else. With nothing to impede the flow, the heat given off by your digesting food would escape rapidly through your skin. If the temperature around you fell too far below your body temperature, you would slowly but inexorably die.

Fortunately, however, your forebears discovered why the skins of the animals you kill are covered with fur. Over time your forebears learned to remove an animal skin with a chipped edged stone, dry it, cut it into suitable shapes and use it as its original possessor did, as body covering to keep you warm. This lifesaving skill has come down to you and your tribe. You even use a flaked pointed stone to cut holes in the skin, through which you can thread narrow strips of other skin, stitching garments and footwear that cling snugly and conserve your precious body-heat.

Sleeping is a further challenge. Even in summer the floor of the cave is cold; so are the walls. You need at least two skins, as furry as possible, to lie on and wrap around yourself - and more in the winter, as the wind whips across the mouth of the cave. Eventually, you or your descendants will hit on the possibility of artificial shelter, easier to warm up and close in than a cave -

11

perhaps a hut of tree branches and skins. The walls of the hut will keep your own body warmth and that of your family from escaping so swiftly to the surroundings.

In the cave you have a fire, close to the mouth of the cave. You don't know where it came from. It has been there since you were tiny, when you accidentally burned yourself on a bright shiny bit of something you found near the fire. Since then you have treated the fire with great respect. You and the rest of the tribe feed it steadily, with the dry wood you drag back from the surrounding forest. The fire eats the wood rapidly, too rapidly. You have to keep finding more. Your elders have taught you well. You must not let the fire die, because no one knows when you might find another, from a lucky lightning strike, without too much rain afterwards. The fire gives you light after sunset, and helps to keep fierce animals away from the cave. It also gives some warmth, if you sit close enough. But it is not much comfort when you're trying to sleep. You can't build it up enough to warm the cave walls. That's just as well. If you did it would probably burn you to death.

Tomorrow six of the fit members of the tribe will set off to hunt mammoth with hand-held spears. It is a dangerous enterprise. A wounded mammoth could injure or even kill them. But if the hunting party succeeds in bringing down a mammoth, the hunters will use flaked-stone hand axes to hack off as much meat as the party can carry back to the cave. It's hard work. Their tools are clumsy, and the meat is heavy. They'll have to leave a lot of meat behind. But that doesn't matter because it would rot before you could eat it all anyway.

Nevertheless mammoth meat keeps you all active and vigorous, able to do the work of hunting and returning with the prey. It will be even better once you discover that you can use the fire to cook the meat, to soften it so that even the elderly and the youngsters can easily chew and swallow it. Moreover, by cooking your food you may find that you no longer have to spend so much time and effort eating. Indeed, over time the brains of your human successors will evolve to become steadily larger and more active, probably at least in part because fire enables cooking and better nutrition.

You manage well enough, keeping warm and doing work, to survive, perhaps even thrive, through day and night, summer and winter. Clothing and shelter, things you have learned to make and use, are crucially important to help you in your daily activities, to help you endure, no matter what the weather or the temperature outside. What you learn from your forebears and pass on to your children, practical experience of what works and what doesn't, especially about how to keep warm, will be a key to survival not only of your tribe but of your species. It will be even more important for your eventual successor species, *Homo sapiens*, when we arrive on the scene.

Now

I am not a Neanderthal. I am writing this on a netbook computer plugged into the battery bank of a solar electric system in a small house on a remote Greek island in the north Aegean. The house is heavily insulated, walls, ceilings and floors, not only against this chill April spring in northern Greece but also against the ferocious heat of the summer to come. I have lighted three oil lamps and a candle, rather than turning on the compact fluorescent lights, because the soft warm glow of the lamps and candles on the white interior walls and ceiling creates an ambience I find gently congenial, especially in the stillness here, the utter absence of motor noise, not even from afar, one of the many reasons I love this place. Shortly I shall open a Mythos, my favourite Greek beer, from the propane fridge.

When I awoke this morning I threw back the fluffy blanket and pulled on my pajama tops and bottoms, and the hooded dressing gown that would make me look like a Mount Athos monk if it were not purple. I made myself a Thermos jug of hot drinking water on the burner of the propane cooker. Then I used a match and a lump of firelighter to ignite the paper and twigs on the grate at the bottom of the old black pot-bellied stove. When the larger logs caught fire the stove leaked smoke again, briefly. Ianni and I plan to seal the joints with fireclay. Ianni, my invaluable island friend, is also going to dismantle the stovepipe to clean it. He says he'll lay the sections on the ground beside the house, and set fire to oily rags inside them.

I sat outside on the terrace watching the sun rise over the hillside to the east, listening to the birds, feeling the surging warmth on my face, and watching the red line on the thermometer in the shade below the honeysuckle creep up from 10 Celsius to 20 Celsius in half an hour. Then I drove the dusty Nissan to the village, noting that the petrol gauge is now below half-full. The only petrol station on the island closed last winter. I shall have to ask Ianni to get me a tankful on the mainland. Although I arrived late at the bakery, I was in time to get the last hot pita fresh from the big oven - 'traditional wood-burning oven', as the hand-painted sign in Greek outside proclaimed.

This afternoon I took a shower, being careful to adjust the taps to avoid being scalded by the water from the solar thermal water heater beside the house. Then I began the regular routine of readying to depart. I ran the generator to pump rainwater from the downhill concrete reservoirs that collect the rainwater from the house roof up to the header tank on the hillside above the house. The generator has enough petrol for at least one more run. I phoned Theodoros, my unfailingly reliable and terrifying Greek taxi-driver friend, to confirm arrangements. Tomorrow morning I shall catch the 0915 ferry to the mainland, to find Theodoros waiting for me on the dock. Flying low in his trusty taxi along the twisting mountain road, we should reach Thessaloniki airport well ahead of time for my return flight to London. Tomorrow evening, all being well, I'll once again be sleeping under another snug blanket in my other comfy bed, in my other cosy home, 1500 kilometers northwest of here.

What we do

I have access to things that even my recent *Homo sapiens* ancestors, never mind Neanderthals, would regard, awestruck, as magic. To be honest I have to admit that I don't know how a lot of them work either. But I can describe what these things do, physically, in only a few words. Many things, in some ways the most important, simply keep heat in or out, or slow its flow - such as the blankets, pajamas, dressing gown and especially the insulated house. They keep me comfortable. Some things produce light. They may do so directly with fire, as even Neanderthals would recognize, such as the oil lamps and candles. Alternatively, they may do so indirectly, not with fire but with electricity, such as the fluorescent lamps with the battery and solar panel. Some things raise temperatures well above human body temperature. The stove, cooker and bakery oven use fire; the solar water heater does not. Some things lower temperature. My fridge keeps my beer cool using fire, remarkably enough, burning propane. It would also run on solar electricity rather than fire, if I so desired.

Some things do physical work for me, that I would otherwise have to do with my muscles, if indeed I could. Without the generator and pump I cannot imagine how long I would take to lift a tonne of water 15 metres, or how I would feel afterwards. The generator uses fire to do the work, burning petrol to make electricity, to run the motor in the pump to produce the force that lifts the water for me. I also have solar electricity, but not enough to run the pump, since we decided to keep the solar array small.

I could walk from my house to the village; but I could not swim the kilometer to the mainland. Walking to Thessaloniki over the mountains would take me weeks, at least. Walking to the Channel coast would take me, realistically, years; and then I would have to swim the Channel. Instead I use fire. The Nissan engine uses fire, burning petrol to move me to the dock. The ferryboat engine uses fire, burning diesel to carry me to the mainland. Theodoros's taxi engine uses fire, burning petrol to carry me at hair-raising speed on the curving highroad over the mountains to Thessaloniki. The plane from Thessaloniki uses fire, burning jet fuel - many tonnes of jet fuel - to carry me and my fellow passengers to London Gatwick airport. Yet another taxi uses fire, burning petrol, to take me the last 80 kilometers to my home outside London.

I have done, am doing and will do all of these things without really thinking about them. But if I thought about them, I'd be thinking about everything we humans do - everything physical, at any rate. The insulated house, blankets, pajamas and dressing gown are all ways of keeping warm - controlling heat-flow. The stove, cooker, water heater, bakery oven and fridge raise and lower temperatures. Lamps, candle and fluorescents make light. The generator and pump exert enough force and do enough work to lift a tonne of water 15 meters in 20 minutes. The Nissan, the ferryboat, Theodoros's taxi and the plane to London move me - give me mobility far beyond what my legs can manage. The solar electric system, phone and computer help me manage information with effortless ease.

Everything I did, and everything we humans do, any time, anywhere, can be summed up in a few words. In physical terms, every human activity you can think of, everything humans have ever done, falls into just six categories. First of all, we control heat flow: for instance, you put on a sweater to keep yourself warm, or open the window to let the heat out. Second, we adjust local temperature: for instance you turn the thermostat up or down. Third, we make light: for instance you light a candle, or switch on an electric lamp. Fourth, we exert force: for instance you lift a weight, or open a door. Fifth, we move things - by exerting force, but moving things is so important you can call it a distinct activity: for instance you push a pram, or pull a wagon. Sixth and last, but in some ways becoming the most important, we manage information: for instance you talk and listen, now not only in person but more and more remotely, with more and more ingenious devices.

Everything we do involves one or more of these six physical activities. That may seem surprising, but it's true. Most of these activities, especially those the most important for survival, we have being doing since the emergence of *Homo sapiens.* So indeed did our Neanderthal precursors before us. Each physical activity started as a human acting alone - if a human alone could do it. A human alone, for instance, could exert force and move objects, using muscles. But a human alone could exert only limited force, and move only moderately small objects. Instead, in almost all our activities we used, and still use, things to help us - physical things. Neanderthals used things they found, such as sticks and stones. But they also learned to use things they made, such as chipped stone axes and knives. We call them tools.

We *Homo sapiens* now mostly use things we have made, 'artefacts', often by extraordinarily long and complex chains of the six basic human activities. In urban areas, especially in rich countries, we are now surrounded by our own artefacts, often far removed from the raw natural resources we used to make them. We now take these things, such as buildings and their contents, as given. We also take for granted their essential roles in our physical activities. When we adjust temperatures, make light, exert forces, move objects and manage information, we do it with the things we have made - or, more likely, with things someone else has made for us. We use them, but we may hardly notice them.

As well as things, we now also use two processes - fire and electricity. As with the things, we use these processes, but hardly notice them. Moreover, most of the time, at least in rich countries, we ourselves may have little control over either fire or electricity. Someone else organizes, provides and controls the process for us. For the past two centuries, ever since we learned to produce and control electricity, electricity has been steadily supplanting fire, in more and more of these activities. We mostly take that, too, for granted. We fail to recognize its importance.

Survival and comfort: controlling heat flow

Have you ever been cold? Really cold, with nothing to keep you warm? We fortunate ones in rich countries rarely if ever experience real cold, and then only because we choose to - skiing, perhaps, or mountain climbing, or Arctic tourism. When we do, we take care to equip ourselves with suitably warm clothing, and try not to stray too far from shelter. But the temperature does not have to be below freezing to be threatening - not if you have no warm clothing and have lost your shelter, as happens all too frequently in earthquakes, tsunamis, floods and wars, even in otherwise wealthy parts of the world. Your survival may depend on finding some way to keep your body from losing its precious heat so fast that your brain and other organs cease to function. In disasters, the most urgent requirement for emergency assistance is almost always to provide blankets and tents - portable, versatile and immediate clothing and shelter to help survivors maintain their body temperature.

Blankets and tents, like all clothing and shelter, function because of the ways heat moves. You would die sooner in cold water than cold air. The heat from your body flows rapidly through your skin into the colder water by what scientists call 'conduction'. Direct contact between something warmer and something colder allows heat to flow across the boundary from warm to cold - the bigger the temperature difference the faster the flow. Water is denser than air, and drains away your body heat more drastically. A diver's wet suit interposes a barrier between your skin and the water, a barrier that does not conduct heat so easily.

If you are naked in cold air, your body heat warms the air next to your skin. It expands and gets less dense. Colder denser air then sinks and pushes the warm air upwards, away from your body, replacing it with a fresh layer of cold air ready to drain off more of your body heat. Scientists call this 'convection'. The process is not as severe as conduction. You don't need a wet suit to stop it. A blanket may do.

You also lose heat a third way. Just as the sun gives off sunlight, and the earth gives off 'earthlight', you give off 'youlight' - so-called 'radiant heat'. Someone with an infrared camera could take a picture in what you think is pitch darkness, and get a vivid image of you apparently glowing in the dark. A shiny surface such as a mirror reflects radiant heat. That's why emergency workers sometimes wrap a survivor in aluminium foil, to keep the survivor's body from radiating away valuable heat.

Sunlight, of course, keeps the planet almost warm enough for all living things. Whether it is quite warm enough by itself depends on where you are and what time of year it is. Bright sunlight in midwinter on a snow-capped mountain is undoubtedly pleasing;

but without clothing and eventual shelter you would still freeze to death.

Heat always moves from somewhere warm to somewhere cooler, if you allow it to. That's why, in most parts of the world most of the time, you wear clothing - the physical reason, nothing to do with style or fashion. Unless you are in extreme conditions in a tropic or desert climate your surroundings are cooler than you are, ready to drain off your body heat. Even when you are concerned not with survival but simply with comfort, you rely on clothing and shelter. Clothing next to your skin is the first barrier to keep you from losing heat too fast. Shelter, some form of structure within which you can move and act, is the second.

Clothing and shelter work together. If you have enough clothing you can manage without shelter, at least for a time. If you have enough shelter you can manage without clothing. But they are more effective in combination. Clothing is personal. It manages your body heat versus all your surroundings, as well as it can; and you yourself can choose how much or how little you need. You know when your clothing is not doing its job. You feel cold or hot, and you can add or remove clothing accordingly. Shelter is collective. If it is large enough, a shelter-structure - a building - can manage the combined body-heat requirements of many thousands of people, keeping them at least tolerably comfortable. However, you also know if the building is not doing its job. You may feel cold or hot, or drafty or clammy; but you may be unable to do anything about it.

To control heat flow, for survival and then for comfort, we have always used physical things. For clothing we use flexible skins and woven fibres, in fabrics and textiles. For shelter, for buildings, we use solid materials - branches and skins, clay, stone, timber, brick, concrete, steel, glass, even blocks of snow - to surround ourselves with barriers and boundaries, to keep heat inside when outdoors is too cold, or to keep heat outside when outdoors is too hot. Such built shelter has enabled us humans to

20

spread across almost the whole of the planet, regardless of the prevailing weather and temperature. In previous centuries we became very good at building effective shelter whatever the surroundings, whatever the materials available.

What do we want from a building? Ever since humans started making structures, buildings have given us many services - shelter, defence, privacy, storage for food, protection for livestock and so on. However, much the most important function of most buildings has always been and still is to provide comfort for those inside. That means, first of all, maintaining a comfortable indoor temperature. That means managing heat. Even when you are asleep your normal healthy body is probably producing about as much heat as a traditional 100-watt light bulb. When you are active your heat output increases. An athlete going flat-out may produce five bulb's-worth. But your healthy body wants to stay at 37 Celsius, 98 Fahrenheit. If you don't want to get steadily hotter you must shed that heat, more or less as fast as you produce it. For most people an indoor temperature close to 20 Celsius, 68 Fahrenheit, usually feels reasonably comfortable. If you are wearing normal indoor clothing inside a building at 20 Celsius, 68 Fahrenheit, the difference in temperature between you and your surroundings is just about right, taking away your excess heat but not actually making you uncomfortably cold.

How does a building maintain a comfortable indoor temperature? First of all it encloses the interior space, reducing the air movement around you when you are indoors. It also impedes both incoming and outgoing radiant heat, depending on how much wall area is windows, and whether the windows transmit or block the radiation. Depending on the design and the materials used in walls, floors, and ceilings, what architects call the 'building envelope', the building structure may conduct heat rapidly or only very slowly in or out. The building envelope also serves as a heat store. When the surroundings are warm the building envelope soaks up heat. Conversely, when the surroundings are cold the building envelope releases heat. It therefore acts as a sort of buffer, keeping temperatures inside more moderate than those outside.

We also use heat indoors. Our most important source of heat is ourselves, digesting the food we eat. Within the shelter of a building, other humans and possibly other animals nearby may help to keep us warm enough. But if the building itself does not keep us adequately comfortable, we resort to other sources of heat.

Fire makes heat at a much higher temperature - much too hot for you to touch without burning yourself. That makes it both useful and dangerous. Fire can keep you warm, but you will be much more comfortable if the fire is not too close, and if you also have clothing and shelter. As a source of heat for comfort, fire works best when we combine it with suitable things. Neanderthals, so far as we know, used fire only by itself - an open bonfire, perhaps in a clearing or the mouth of a cave. Now, however, we *Homo sapiens* use fire in countless ways, in combination with countless things. We now use fire much more than we used to. Most of the time we don't even realize we are using it.

We also have one further source of heat, that humans have used for less than two centuries - electricity. Some electricity we produce using fire, usually by burning coal, oil or natural gas. Some we produce from an even more violent extreme process, nuclear fission in uranium. Some electricity, however, we produce without using fire or the high temperature it creates. Instead we use physical things that respond to the forces of nature - wind, sunlight, falling water, waves and tides, and even the heat from inside the earth - to produce useful electricity. This electricity in turn will also produce heat, at a temperature that can be high but need not be.

Electricity also enables us to sense our surroundings, in ways that let us use the information to control heat flow, temperature and other factors affecting our comfort. As yet we have barely tapped the potential this offers. To use electricity for control rather than fire for heat - the battle is just beginning.

Heating things

Have you had hot food today? How did you heat it - with fire or with electricity? For cooking, we have had the electric option for less than a century. Cooking with fire, however, may have been a key step in the early evolution of the human species - so some scholars now plausibly suggest. The raw meat that was the main food for Neanderthals and early *Homo sapiens* required strong healthy teeth and a robust digestion. Heating the meat with fire - cooking it - would soften it, making it easier for the young and the elderly to chew and for everyone to eat and digest quickly, a significant advantage for the survival of the tribe and eventually for the evolution of human beings.

To raise local temperature and heat things we have always used fire. We still do, especially for what has long been the universal human activity of cooking. In rich countries, to cook with fire, when we have access to natural gas we now commonly prefer it for cooking, rather than the wood or coal fire common a century ago. Whatever we burn, we also take care to ventilate the kitchen to remove the combustion products. In poor rural areas, however, especially in Africa and Asia, women do the cooking. First they have to find and gather firewood or dung, often requiring several hours on foot, as feeding fire has stripped the countryside of trees. Then they light the fire, usually indoors, perhaps between three stones under the cooking pot, devouring the firewood all too fast and filling the house with smoke. Inhaling the resulting smoke and fumes is a major cause of illness and death, for both women and children.

In rich countries we have been able to cook without fire for nearly a century. When electricity flows through a wire, if the wire is thin enough or the electric current strong enough, the wire may heat up. Such hot wires can be used simply to heat a room, with what in the UK is called, paradoxically, an 'electric fire'. They can also be used on top of an electric cooker, or inside it, as an electric oven. More recently, a very different way to deliver heat electrically has also become commonplace, the so-called 'microwave oven'. It delivers a form of radiant heat akin to X-rays, that penetrates the food placed in the oven, effectively heating it from the inside out. Electric heating in its various forms is generally easier to control than heating with fire.

Cooking, the intentional use of fire to process food, might also have led to a further important development. Intentional heating with fire could harden clay into ceramics for vessels, watertight and durable, and eventually for materials such as tiles and bricks for sturdy watertight shelter. Ceramic vessels, in turn, made cooking easier, and offered the possibility of cooking not only over a direct flame but also in heated water, widening the range of edible materials and combinations of them. In due course we learned to confine the fire inside a housing called a kiln, to bake ceramic pottery. When we could make the fire hot enough, we could melt a sand coating into glass, making the glazed vessel more watertight and easier to clean. We still use much the same process today.

Somehow, when heating things, we found, perhaps to our surprise, that hot shiny liquid ran out of certain rocks. When it cooled it hardened into firm but slightly soft metal - lead from some rocks, tin from others, and eventually, with hotter fire, copper from still others. We learned to shape these metals by pouring them, molten, into patterns or molds, to make ornaments,

24

vessels and implements. We also learned to mix different molten metals - tin and lead to make pewter, tin and copper to make bronze. These so-called 'alloys' were more suitable for some applications than the pure metals. 'Smelting' with fire freed the metals from their ore. 'Casting' and 'forging' with fire melted or softened them, so that we could make useful metal artefacts.

In due course we found a way to make fire much hotter, using charcoal as fuel. To make charcoal you roast wood without letting in enough air for it to burn. That drives off all the moisture and volatile constituents of the wood, that would keep the temperature lower while they vaporize. The charcoal that is left burns much hotter than wood. With charcoal fire and extra air, perhaps blown by a bellows, we found we could smelt iron metal from its ore. Iron proved to be much more useful than any of the earlier metals. It was much harder and stronger. You could give it an edge or a point that would remain sharp - a valuable attribute for tools, implements and weapons. Moreover, by controlling the fire, you could have some of the carbon from the charcoal combine with the iron to make an even more versatile metal - steel. Over time, understanding the complex behaviour of metals with fire, including careful cooling, became first a craft, then a trade, and eventually the science of metallurgy and a vast global industry.

Heating other things with fire followed a similar progression. Firing ceramics, bricks and other materials, making cement, refining petroleum, manufacturing chemicals and plastics, and other industrial activities requiring high temperatures now use fire as an essential part of the production process. But its consequences cause increasing concern. Meanwhile, some industries are finding that delivering heat by electricity is cleaner, more versatile and easier to control. In industry, electricity is steadily displacing fire. But fire still has a long lead.

Making light: with fire

Making light was almost certainly the most important original human use of fire, dating back to our Neanderthal forerunners. The sun sends us light - but only part of the time in any particular place on earth. Human senses evolved with light. Hearing, touch, smell and taste function anywhere, at any time; but seeing needs light. Fire could make light from sunset to dawn. Our Neanderthal forerunners probably also learned that the brightness of fire in the darkness of night would deter wild animals from attacking them. Once we learned how to control fire, making light became another fundamental human physical activity.

Since then, throughout most of human existence, we have made light with fire: a burning brand, a torch, a cup of plant oil with a floating, burning wick, a candle of animal tallow, all giving off smoky yellow-orange light, tolerable outdoors, less so indoors, coating walls with soot and creating an oppressive fug in the room. Then came oil lamps with glass chimneys, cleaner and brighter, often burning oil from the blubber of whales. Then came gas-light, a major change. Instead of individual lamps that you had to keep filling with fuel, gas-light meant fixed fittings in your building, connected by permanent pipework all the way to the gas plant, possibly kilometers away. From 1800 onwards, gas-light from town gas sprang up all over Europe, North America and elsewhere.

From the mid-1800s onwards, however, the advent of electric light, first arc-light, then the incandescent lamp, rapidly superseded firelight of every kind almost wherever electricity became available. The desire for electric light drove the rapid expansion of electricity systems throughout Europe and North America, and subsequently through Asia and Latin America, particularly in cities. Nevertheless in rural areas, especially in Africa, even now many people still remain without electric light or other services from electricity. Instead they still rely on firelight, notably from kerosene, expensive, smelly and smoky, aggravating indoor breathing problems. When they do have electricity it is often from fire in diesel generators, noisy and smelly, burning fuel that must travel often hundreds of kilometers over poor roads and is accordingly very expensive.

Making light was the first human physical activity in which electricity directly challenged fire. Even for making light, however, fire still dominates. Too many people still have no electricity.

Exerting force: without fire

The force you exert most often you probably hardly notice. Every time you sit up, stand up or pick anything up you are exerting a force to overcome the force of gravity. You do so with your own muscles, as did our Neanderthal precursors. Unlike the Neanderthals, however, we can exert forces in many other ways, not only to overcome gravity but to shape materials and move and manipulate objects. As well as using our own muscles we can use those of animals that we have domesticated for the purpose - horses and donkeys, oxen, elephants, camels, sled dogs and so on. Of course the Egyptians, Greeks, Romans and other ancient societies also relied on the muscles of human slaves, often treating them more brutally than domestic animals. But their combined force was capable, for instance, of raising the vast pyramids in Egypt.

Human and animal muscle strength is limited. In past millennia we invented or discovered ways to multiply that strength. We used other physical things, what the Greeks called 'simple machines', to make tasks easier. A 'lever' such as the crowbar, the wheelbarrow or the fishing rod multiplies or manipulates forces. An 'inclined plane' such as a ramp allows you to move a heavy weight uphill more easily. A 'wheel and axle' helps, for instance, to lift a bucket from a well; you pull down and the bucket comes up. A 'block and tackle' turns a single rope into, say, four; you pull on the single rope and your force is delivered four times; and so on.

Nature also exerts other forces besides that of gravity. A strong wind can knock you over. So can a fast-moving river. In due course we learned to use physical things to capture the forces of moving air and water, to redirect these forces for our human purposes. The ancient Greeks, and subsequently Romans, Indians and Chinese, used wheels spun by moving water - 'water wheels' or 'water mills' - to turn grindstones for grain, to drive hammers and for other applications. The early Chinese, the Phoenicians in the Mediterranean, and in due course many others, propelled their vessels with wind as well as muscles, with sails as well as oars. The Cretans erected what we now call windmills with blades resembling sails, mainly for pumping water. But windmills soon became widespread, also for milling grain and other purposes.

For most of human existence fire was of no use for exerting force. Natural forces and simple machines nevertheless made potent and effective combinations. All of these activities relied not on fire but on harvesting and reorganizing natural phenomena to do what we wanted to do. It was an early foretaste of what may now be happening again, millennia later.

Exerting force: with fire

Although we had long used fire to produce heat and light, until just over 300 years ago no one used fire to exert force. When at last someone did, the basic idea was so simple you have to wonder why it took so long to think of it. When you use fire to boil water, the water turns from liquid to gas - steam. In doing so its volume increases perhaps a thousandfold, depending on how you boil it. If you boil it in a confined space, it wants to expand but cannot. It therefore exerts force on the container - the hotter the steam the more force. If you heat a tin of beans in a campfire, you have to be sure you puncture the tin beforehand, to let the steam escape. Otherwise, the force will build up until the tin explodes - at the very least an embarrassing mess, and potentially lethal.

Hero of Alexandria, a Greek engineer, did indeed notice the potential of this phenomenon. He even made a working model, a hollow sphere spun by jets of steam. But it remained a toy and curiosity. Some 1700 years later, a British engineer called Thomas Newcomen was trying to find a way to pump water out of a flooded coal mine. In 1712 he hit on the notion of using fire to boil water inside a cylinder, with a sliding piston at one end. As the water turned to steam, the pressure moved the piston, exerting enough force to move a beam and lift water up the mineshaft. Newcomen called it an 'atmospheric engine'. It was the first version of what we now call a 'steam engine'. However, to get the piston back to its starting position the Newcomen engine had to let the steam out. Each stroke of the piston had to start by boiling up fresh steam. The fire to boil the water burned coal from the mine. Letting the steam out each time meant that the fire used an alarming amount of coal to lift the water. The Newcomen engine worked, but not very well.

Then, starting in 1763, the Scottish inventor and engineer James Watt, working at the University of Glasgow, came up with a crucial modification of Newcomen's design, dramatically reducing the amount of fuel the steam engine needed to operate. Instead of simply letting the steam out, Watt's design reused it. Pipes and valves channeled steam to one side of the piston until it reached the end of its stroke. The valves then channeled the steam to the other side of the piston to push it back to its starting point. After struggles with finance, Watt teamed up with another engineer called Matthew Boulton, and found commercial success. When the Newcomen engine started to be used in coal mines, it used the 'small coal' from the same mine, which could not otherwise be sold; its performance did not matter much. But Watt's engine improved performance fivefold. Using his engine to pump out water from the tin mines in Cornwall, where the coal had to come by sea, became economic. Boulton and Watt then leased their engines, charging one-third of what the engine saved on coal – a financial innovation looking far into the future. Before 1800 the firm of Boulton & Watt was selling and leasing steam engines in quantity.

Using fire, the steam engine was smoky and smelly, and also very noisy. But being able to use fire to exert force, much more force than you could get from muscles of any kind, and whenever and wherever you wanted it, triggered a dramatic change in what humans did and how they did it. For millennia, artisans and craftspeople had exerted skilled and precise forces with tools to make textiles, clothing, pottery, utensils, furniture and other useful things. They 'manufactured' these things - made them by hand, one at a time. By the late 1700s some entrepreneurs, notably Richard Arkwright in his textile factory, were using machines driven by water power to speed up manufacturing. Using fire in the steam engine then set off a rush to devise more machines to use the force it could exert, to make things far faster than the hands of individual artisans ever could. By 1800 the steam engine was steadily replacing muscles. Manufacturing was no longer manual. The change came to be called the industrial revolution, initially in the UK, then ever farther afield. Exerting force with fire transformed human society. We are still struggling with the consequences.

Moving things: without fire

One of the commonest reasons you exert a force is to move something. Even to move yourself, your muscles are exerting forces on the bones in your arms and legs, and so on. You could argue that, as a human physical activity, moving things is just an example of exerting forces. But moving things, especially in modern human society, is such an important and varied physical activity it deserves a category of its own.

For the Neanderthal hunting party, bringing the fresh mammoth meat back to the cave was a major undertaking, a lot of effort, all of it by their own muscles. In many rural parts of the world human muscles are still the main way to move things, slowly and often painfully. Elsewhere, however, in rich countries and also in the teeming cities of poorer countries, people and goods are continuously in furious movement. Apart from human and animal muscles alone, some use wheels to move - bicycles, rickshaws, wagons, pushcarts. But many now also use fire, in countless engines burning wood, coal, petrol, diesel, kerosene and other fuels, in vehicles of every size and kind, on land and sea and in the air.

They are moving people and things because they have to. The Neanderthal tribe probably stayed close together most of the time.

Except for the occasional hunting party, they had everything they needed close by. Until not much more than a century ago most people lived, worked and played within a neighbourhood they could reach on foot. In the rural areas of poor countries most still do. But many of us, especially in rich countries, have spread out our lives well beyond a day's journey, if we are using only our own muscles. We now live far from the sources of our food and fresh water, far from our work, even far from our leisure. Using fire in vehicle engines, first trains, then cars and trucks, allowed us to spread out our activities, wider and wider. Now we have spread them so widely that most of us in rich countries can no long manage without moving, and counting on others to move, almost every day, distances possible only by using fire - or electricity.

Moving things: with fire

The steam engine first enabled us to use fire to move things. Pumping engines moved water. Stationary traction engines pulled ploughs with cables. But the major breakthrough was to make the fire in the steam engine move itself. The first success was on water. A steam engine mounted in a boat could turn a paddle wheel or a propellor, replacing oars or sails with motive power from fire and fuel. Steamboats became successful particularly on inland waterways, where sails were of limited use and fuel could be gathered from the banks. They were less immediately successful on the open ocean. The wind was free, not so the fuel, which also took up a lot of space that might otherwise be paying cargo.

Waterways already existed. Roadways smooth and firm enough to bear the weight of a moving steam engine did not. But horsedrawn vehicles were already using wooden or steel rails to provide a more level track. In 1829 the Liverpool & Manchester railway held a competition at Rainhill to decide which engineers would win a contract to supply 'locomotives' - self-propelling engines - for the new railway. The 'Rocket', of George and Robert Stephenson, won the Rainhill trials. Soon engineers and entrepreneurs were using steam locomotives to pull trains of wagons carrying people and goods on ever-expanding networks of railways, again initially in the UK but soon much more widely. Fire was becoming the key to increasing mobility on land, far beyond the capability of human and animal muscles. By the end of the nineteenth century fire also overtook oars and sails for moving things on water, even in the open ocean. But moving things with fire was only getting started.

Gas-light: a fire system

Working for Boulton & Watt in the 1790s, the engineer William Murdoch made yet another breakthrough with major long-term implications. Once again improving on the work of several precursors, Murdoch demonstrated that roasting coal with fire inside a closed retort produced a gas that would burn, so-called 'town gas', colourless and poisonous, that burned just like whale oil, except smokier and smellier. He used the fire of gas-light to illuminate his house in Cornwall in 1792, and the company's works in Birmingham in 1798. By 1807 Pall Mall in London had become the first public gas-lit street. Within a decade gas-lighting was spreading across Europe and North America, both on private

33

premises and in public spaces, especially for street lighting, where the smoke and smell were less noticeable.

If you were a gas-lighting entrepreneur, you had a lot to do. You had to design or license the hardware - retorts, pipes, valves and burners. You had to arrange to fabricate and install them, hiring and paying for the necessary skilled workers. You had to find adequate finance for the investment and operating costs; and you had to persuade prospective customers to sign up for the service. Since the customers might be some distance, even a kilometer or more, away from the retort, the pipeline had to pass through space belonging neither to your company nor to the customer. To lay the pipeline you therefore had to get permission from whatever local government oversaw public space. You then had to maintain the pipeline network, not only for reasons of business but also for safety; a gas leak could lead to fire or explosion. Your company had thenceforth a direct physical connection to the premises of each of your customers, unlike any of the previous forms of business supporting human physical activities. You fed each customer's fire continuously, from a distance.

You had to purchase a continuing supply of coal for the retorts, and to dispose of the consequent solid waste. You had to control the production process to ensure a reasonably consistent composition of the fuel gas fed into the pipework, so that it would ignite and burn with a smooth reliable flame at the customers' gas lamps. This was an early manifestation of what was to become of critical importance, the ever more stringent specification for the fuel to match what you burned it in. You had to create and operate an entire system, physical, financial and managerial, to sustain the fire of gas-light. You might also have to comply with whatever constraints the government might impose, as the notion of government regulation of such activities took root.

Electricity: a new process

Throughout the 1800s, while the possible uses of fire were ramifying, another process was making its potential ever more evident. The ancient Greeks knew that if you rubbed a piece of amber, a hard translucent resin from trees, with a piece of fur, the amber would behave oddly, attracting scraps of parchment. The Greek word for amber is 'elektron'. The Greeks also knew that certain types of rock exhibited a similar behaviour, attracting iron nails. By 1600 scientists had described what came to be called 'electricity' and 'magnetism'. But a further two centuries passed before they began to yield practical applications. Then, between 1790 and 1840, electricity at last moved beyond a party trick. Scientists created the electric battery and electric current and recognized that electricity and magnetism were closely interrelated. Building on the work of Volta, Oersted and others, Michael Faraday demonstrated that moving a wire in a magnetic field made electric current flow in the wire, and conversely that an electric current in a wire produced a magnetic field. The discoveries led to the invention of the 'dynamo', to generate electricity from motion, and the electric motor, to create motion from electricity.

Electric telegraphy: managing information

You may not think that managing information is a physical activity. But if you think of it, that is what you are doing. We can

35

now scan your brain or mine, and detect the activity that is thinking. It is electrical activity - minutely subtle electric currents flowing through brain tissue, manipulating what your senses tell you about the world around you and what you think about it. As you read these words you are processing information. Your brain scan would show it. Despite many years of research, we still do not really understand how our brains process information. But one thing we do know: your brain processes information at your body temperature. If it goes up even two or three degrees, a fever, you can no longer think straight.

Thinking helps you to organize and respond to what your senses tell you. Thinking prompts and directs much of your other physical activity. For us humans, one profoundly important manifestation of thinking is language: 'I don't know what I think until I hear what I say.' Language gives you a clearer grasp of your own thoughts. It also allows you to express yourself, to communicate your thoughts to others, provided they also use a similar language. Language can take many forms, including speech, music, visual art and writing. Language also makes possible counting, measuring and arithmetic. Language links information and communication.

Humans, like their great-ape cousins, have always been social and gregarious. That requires communication. The Neanderthals had to process and share information to plan the mammoth hunt - who would be in the hunting party, which direction to go, what provisions to take, who would carry the spears and hand-axes, who would carry stitched-skin sacks for the meat and so on. *Homo sapiens* developed information and communication much further. Cave paintings close to 40 000 years old are vivid descriptions of hunting, hunters, their weapons and their prey. Archaeologists have also found musical instruments, such as bone flutes, that may be almost as old.

Written records are more recent, initially arising most commonly in commerce - bills of sale, lists of goods and the like. As society became more layered, with the emergence of an elite, kings and priests and government, writing became a necessity for administering taxes, census-taking and other functions. It also became a way to acclaim and magnify the exploits and accomplishments of those in power, as did visual art, painting and sculpture. Scholars used to assert that language was a uniquely human achievement, that only humans used language. We now know better. Depending on what you mean by language, many other animals also use it, from great whales to honeybees. But we can identify one other achievement that does seem to be uniquely human. Not even the great apes can start a fire.

Information and communication became a hallmark of human society. Fire, however, had little to do with it. Firelight made writing and reading, for those few who could do so, a little easier. Once in a while more dramatic fire conveyed important news, such as the English beacons warning of the Spanish armada. But fire's most important contribution to managing information came from making possible the use of electricity.

The first really successful long-distance communication faster than beacon fires or a stagecoach was the electric telegraph, the first practical use of electricity. In the 1830s, William Cooke and Charles Wheatstone in Britain and Samuel Morse in the US showed that by sending electric current along a wire you could transmit a message along the wire. Morse's method, starting and stopping the electric current according to the code he devised, proved the simplest and most effective. A skilled operator could send information ever greater distances almost instantaneously - 'electric telegraphy'. Within a few years telegraph wires and cables were unrolling at breakneck speed, first across and then between continents, over mountains and under oceans. By the beginning of the 1900s the electric telegraph connected much of the world, in what we now call real time. The electricity it used, at least in its early stages, came not from fire but from chemical batteries - an early indication that fire was no longer the only process we humans could use in our activities. Making the copper for wires and cables, however, did use fire.

Arc-light: light without fire

Even as the electric telegraph was demonstrating the potential of electricity for managing information, inventors were devising other ways to use electricity in human activities. Their next success was for making light. By the 1860s scientists and engineers had succeeded in creating artificial lightning - a continuous electric spark hissing across a gap between two sticks of carbon called 'electrodes', making garishly brilliant bluewhite 'arc-light'. It was noisy and smelly and fiercely bright, far brighter than any form of firelight, utterly impractical for lighting indoors. But it was spectacular display lighting outdoors.

Arc-light, like gas-light, required not just a lamp and something to burn in it, but an entire elaborate system of hardware. Electricity entrepreneurs persuaded wealthy clients to install complete electricity systems, including dynamo, cables, switches, arc-lamps and controls. To make the electricity you had to spin a coil of wire between magnets. To spin the coil you might use, say, the mill wheel in a water mill. That would work only if you wanted the arc light somewhere near the water mill. Another option would be to spin the coil with a steam engine. That you could do anywhere, provided you could get something for the steam engine to burn. Using a steam engine to drive arc-lighting neatly inverted the original phenomenon. Instead of getting fire from electricity, like the Neanderthals, you were getting electricity from fire. On the other hand, if you used a water mill, you were generating electricity not from fire but from natural forces. Indeed your electricity was therefore a well-behaved first cousin of nature's electricity - lightning.

Fire or lightning? - for the electricity we use, the distinction has become crucial.

Incandescent light: an electricity system

Arc-light had some of the same problems as gas-light. Like gas-light, arc-light was unpleasant indoors. Gas-light was dim, much fainter than daylight, hard to read by; but arc-light was harsh, almost painfully bright, even worse in a room than gas-light. Then, at the end of the 1870s, Thomas Edison in the US and Joseph Swan in the UK came up with a dramatic improvement for electric light - the incandescent lamp. Sending electricity through a thin filament inside a glass bulb with no air in it made the filament glow, a warm, silent and odorless light far brighter than gas-light and far more congenial than arc-light, eminently suitable for lighting indoors.

The incandescent lamp also allowed Edison to seize another opportunity. Because it made electricity flow through a very thin filament, you could wire up many incandescent lamps effectively side by side, in 'parallel', dividing up the current flow between them. You could therefore make the whole system much larger, with much larger total current. That offered an important improvement. You could double the size of a dynamo or a steam engine without doubling its cost. Making the electric light system much larger was therefore a way to reduce the overall cost to each customer - essential if electric light were ever to be more than an ostentatious display for the wealthiest.

In 1882, Edison set up an incandescent-light system in lower Manhattan, supplying illumination to Wall Street and other suitably wealthy premises nearby, from a generating station in Pearl Street. Like gas-light entrepreneurs before him he had to get permission to use the public streets, to run cables from the generator to his customers. He had to supply and install not only the dynamo and the steam engine to run it but also the cables, the switches and the lamps - the whole system. At the outset he charged his customers according to how many lamps they had. He was selling access to illumination - exactly what his customers wanted. To keep the cost to the customers from being even more daunting, he had to optimize the entire system - every component, including the lamps.

Then, in 1885, came a practical electricity meter, able to measure how much electricity any given customer used. Suddenly Edison and his fellow electricity entrepreneurs were no longer selling illumination. They were selling electricity, by the unit, as measured by the meter. Suddenly they no longer wanted to optimize the entire system - on the contrary. A customer using ineffective lamps had to buy more electricity to get the same illumination. From that time on, companies selling electricity actually benefitted from having customers use ineffective lamps - a perverse incentive that still applies today.

Edison's systems for arc light and then for incandescent light used electricity like that from a battery, in which the current always flowed the same way through the wire - so-called 'direct current' or DC. Edison's DC faced one major problem. Current flowing in a wire under a certain electrical pressure or 'voltage' made the wire hot - the more current the hotter. The heat escaped, weakening the useful effect of the electricity. The longer the wire, the worse were the losses. You could not, therefore, extend your network of wires very far from the generator. That in turn limited the potential size and reach of your system. It was a problem if you used fire, in a steam engine, to turn the dynamo. It was an even bigger problem if you wanted to use, say, a water wheel, which had to be where the water was, not where you might want lamps.

To offset the problem of line losses, you could use thicker wire. But thicker wire rapidly grew much more expensive, especially for long lines. You did have one other possibility. To get the same electrical effect in the wire you could double the voltage and halve the current. The heating effect would then drop to a quarter, reducing the losses from the wire. With DC, however, you had no way to double the voltage and halve the current.

Even as Edison was establishing his first large-scale DC systems in the early 1880s, a scientist-engineer called Nicola Tesla and his colleague George Westinghouse began promoting a different system, taking advantage of the rotation of a dynamo. Successive half-turns of the rotation made the current in the wire surge rapidly back and forth - so-called 'alternating current' or AC. A dynamo producing AC became known as an 'alternator'. Its speed of rotation determined how many times per second the current surged back and forth - its 'frequency'. With AC you could use a device called a 'transformer'. It would increase voltage and decrease current, or vice versa, depending on how you wired it. By using AC and transformers you could send electricity much farther with much lower losses. From the alternator you sent the electricity through a 'step-up' transformer to raise its voltage as high as possible, and reduce the current accordingly. The smaller current heated the wire much less. Then you sent the electricity through a 'step-down' transformer, to restore the original voltage and current, before sending it through the filaments of the incandescent lamps, heating them up to a bright glow as desired.

AC was so successful that by 1895 entrepreneurs had set up a generating station upstream of Niagara Falls, using the rushing water to turn an alternator and send so-called 'hydroelectricity' all the way to Buffalo, New York, to light incandescent lamps more than 50 kilometers away. By making long-distance 'transmission' of electricity feasible and practical, AC removed the distance constraint that had previously limited the use of natural forces to generate electricity. From then on using fire to make electricity faced much more significant competition from natural forces.

Exerting force: with electricity

Incandescent electric light soon supplanted firelight of every kind wherever you could access electricity; and the desire for electric light dramatically accelerated the spread of electricity systems, in Europe and North America and then farther afield. The new electricity systems, however, faced one obvious difficulty: because electricity is not a substance but a process, you must generate it immediately as it is used, and vice versa. People wanted electric light after sunset and, more or less, before midnight - only a few hours a day. For the rest of the time the expensive investment in dynamos, cables and other hardware was idle, its owners probably paying interest on the investment but earning no revenue. Electricity entrepreneurs looked for a way for customers to use electricity in the daytime. The answer was the electric motor.

Just as a dynamo changed motion into electricity, an electric motor turned electricity into motion, exerting force to do so. Factories using steam engines to exert force found that electric motors have major advantages. Electric motors are not exactly quiet, but much less noisy than steam engines. Electric motors are not smoky or smelly; they do not require fuel bunkers, often dirty and intrusive, on the premises; and they do not produce ash needing disposal. They start up and shut down almost instantly. They are also easier to control, over a wider range of sizes. From the viewpoint of electricity entrepreneurs, factories became valuable customers, using electric motors in the daytime when few people turned on lamps. Even before the beginning of the twentieth century, both for making light with incandescent lamps and for exerting force with electric motors, electricity rapidly began replacing the direct use of fire.

The first electric motors, designs inspired directly by Faraday's research, ran only on DC. Then Tesla invented an AC motor, whose performance and potential soon exceeded that of the available DC motors, another advantage for AC. Edison, however, stoutly defended DC against the mounting challenge from AC. He declared that AC was more dangerous than DC, a claim disingenuous at best. In a bizarre and ugly episode he even promoted the use of AC in the 'electric chair', to execute criminals, as a demonstration of its danger. The 'battle of the systems', DC versus AC, raged into the early 1890s. Then General Electric, Edison's own company, conceded defeat. It eased Edison out the door and embraced AC. But DC did not disappear. In due course it was going to make a striking comeback.

Moving things: with electricity

Electricity also began to play a role moving things. Just as the steam engine could move itself, so could the electric motor, as long as it could be connected to an electricity generator. Companies providing public transit in urban areas with horsedrawn trams found that they could introduce electric trams, drawing electricity from overhead wires. Electric trams were so successful that tram companies began installing their own generators, rather to the annoyance of the incumbent electricity companies, who wanted the tram companies as daytime customers.

Inventors also began efforts to power 'horseless carriages' with electric motors. For many years an assortment of short-lived and unsuccessful designs went nowhere. For a vehicle, you had to carry your electricity generator with you. That meant a battery. Inventors tried various designs of battery, but they all ran down too fast. The invention of the lead-acid battery, which could be repeatedly recharged, was a major breakthrough. For the first time you had a way to 'store' electricity made elsewhere. Of course you were not storing electricity itself. Electricity is a process, something happening. You cannot store something happening. But you can store what makes it happen. In a rechargeable battery you are storing chemicals that will give you back the electricity when you want it. By the late 1890s electric vehicles including bicycles, tricycles, cars and taxis running on rechargeable batteries were to be seen in the UK, France, Germany, the US and elsewhere. Some of the electricity stored in these vehicle batteries came from generators using fire in steam engines. Some, on the other hand, came from generators producing hydroelectricity - an early indication that you could now move things with machines without using fire.

Fire inside

In the steam engine, the fire was outside the cylinder in which the piston moved. By the mid-1800s engineers were seeking ways to ignite the fire inside the cylinder, to move the piston not with steam but with hot expanding gas from the fire itself - 'internal combustion'. Initial success came using as fuel the coal gas produced to make gas light. You injected the coal gas into the cylinder and exploded it there. To make it explode, however, you did not use fire. Instead the engine incorporated a device that created an electric spark inside the cylinder, setting off the gas explosion. Gas engines began to compete with steam engines in factories.

A gas engine of course required a permanent connection to the retort producing the coal gas. But another option soon emerged. Beginning in the mid-1800s, entrepreneurs began paying serious attention to seepages of oil emerging from the ground in places including Romania and the northeastern US - oil that would burn, rock oil or 'petroleum'. Within a few years, drillers were producing petroleum in quantity to replace whale oil for making light. By distilling or 'refining' the crude petroleum they got a fraction called 'kerosene' that burned more reliably in oil lamps. Wherever kerosene was available it rapidly took the place of whale oil, a rapidly expanding market for the petroleum refiners.

Refining also produced another fraction, that Europeans called 'petrol', 'benzin', 'essence' and other names, and Americans 'gasoline'. At first it was a surplus product that no one particularly wanted. It was dangerously volatile to use in an oil lamp. But it proved to be ideal to use in an internal combustion engine. You could have a tank of liquid fuel, petrol or gasoline, beside the engine. When you injected it into the cylinder it vaporized, so that a spark would explode it, thrusting the piston to exert a force to turn a shaft. Moreover you could carry the tank of fuel on a vehicle. Yet another fraction of petroleum also proved to be similarly useful. The German inventor Rudolf Diesel demonstrated an engine in which this fraction, subsequently named after him, would explode in the cylinder without needing a spark. You simply had to compress it sharply with the piston, which raised its temperature enough for it to detonate.

By 1885 the internal combustion engine, in both petrol and diesel versions, had become yet another way to use fire, both to exert force and to move things.

Moving things: fire versus electricity

Throughout the nineteenth century more and more of us wanted to move more and more often, faster and farther. Fire expanded our options, in steamboats and railways, and subsequently in road vehicles, first with steam engines and then with internal combustion engines. Steam road vehicles were initially so heavy they could only be used on the rare well-made roads. They were also slow to start, because you had to ignite a fire and raise steam. You could start an internal combustion engine quickly, if you were strong enough, but you probably had to crank it repeatedly before the sparks ignited the fuel and began moving the pistons. Electric vehicles, also in the race and indeed for a time leading it, did not have the problem of starting a fire. But they depended on batteries for their electricity, and the available batteries ran down too quickly, limiting vehicle range and speed. You could use more batteries, but they were heavy and expensive. Nevertheless, by the end of the nineteenth century electric vehicles and those using fire by steam or internal combustion were all competing for the custom of the small number of adequately wealthy drivers, mostly in Europe and North America.

Ironically, electricity helped the internal combustion engine to beat the electric car. The invention of the self-starter, a battery-powered electric motor that replaced the hand-crank to turn the engine over, removed one of the main impediments hitherto holding back the internal-combustion car. But the largest advantage of internal-combustion fire over electricity was one taken completely for granted at the time. We allowed the internal combustion engine, burning petrol or diesel fuel, to emit its noxious exhaust gases directly, copiously and freely into the public air.

At the beginning of the twentieth century all three types of car were still rare luxury items, available only to the wealthy; but they were desired by the many watching on the roadsides as they passed. Then the American entrepreneur Henry Ford invented the assembly line. Mass production dramatically reduced the cost of a motor car, and brought car-ownership to ordinary people. The motive power Ford chose was the internal combustion engine. The marriage of cheap petrol and the fire of internal combustion changed the shape of human society worldwide.

Fire at sea

While making steady headway on land, with steam railways, steam and internal combustion cars, and electric trams powered by steam-generated electricity, fire was only slowly overtaking wind for moving vessels at sea. By the beginning of the twentieth century steamboats were plying inland waterways and along coasts, but only gradually taking to the deep sea. When land was nearby, fuel to feed the fire in the steam engine was readily available. On the open ocean, however, it had to be carried, taking up space and tonnage that could no longer be used for cargo. On the other hand, escaping the vagaries of the wind meant that voyages could be more direct and potentially even swifter, and would not need so many hands to manage the sails, a trade-off that became rapidly more attractive as wages rose and the performance of marine steam engines improved.

Steam power also began to play a role in sea power, as the world's navies gradually abandoned sail for fire. In the early years the fuel for sea-going steam engines was usually coal, as it was on land. Then, in 1911, Winston Churchill, as the UK government minister responsible for the Royal Navy, opted to switch from coal to oil to feed the fires in its ships. The decision was a major boost to the petroleum industry, opening a huge new market for the heavy fraction burned in marine steam engines. That also meant an upsurge in production of the lighter fractions that included petrol and diesel. Making these fuels more widely available and cheaper further fostered the growing popular enthusiasm for motor vehicles run by fire.

Cooling things

Heating things - adding heat - is easy, especially if you have fire. Cooling things - removing heat - is harder. Left to itself heat always moves from a warmer place to a colder. The easiest way to remove heat is to have a colder place, for instance a block of ice, nearby. Through the millennia, and well into the twentieth century, people with adequate winters learned to harvest and store ice, sometimes in impressive quantities, to provide cooling when they wanted it. Your domestic household might have an 'icebox', which was exactly what it sounds like: an insulated box containing a block of ice, next to which you placed your food and drink to keep it cold and fresh. You had a regular visit every few days from the 'iceman', carrying with a pair of tongs a fresh block of ice, to replace the ice that had melted while removing heat from your food and drink.

You can also cool down by fanning yourself, blowing away the air warmed by your skin and replacing it. But if the air itself is too warm this is not much use. Sweating helps. Fanning your wet skin now blows away air full of water vapour, and replaces it with air ready to aid evaporation. Evaporating water absorbs heat from your skin, cooling it. Evaporation to soak up heat became a key to the first commercially successful cooling process that did not require winter temperatures to make ice.

If you want, say, to cool the interior of a box, to make it a 'refrigerator', you have to make heat move out of the refrigerator, into the warmer room outside it. That is, you have to make heat move from colder to warmer, against its natural tendency. In effect you have to 'pump' the heat 'uphill'. One way to do this is to have a volatile fluid - a 'refrigerant' - in a closed loop. On one side of the loop, outside your refrigerator, a pump compresses the fluid, raising its temperature, so that heat flows out of the loop. If you stick your hand behind your refrigerator you can feel the heat coming off the cooling vanes.

Then the compressed fluid goes through a narrow nozzle into the low-pressure section behind the pump, inside your refrigerator. As the fluid expands into this section some of it evaporates, rapidly cooling the remaining liquid. That in turn absorbs heat from the space inside your refrigerator. Then the pump compresses the fluid again, warming it up enough to dump the heat outside the refrigerator - and so on, in a continuous operating cycle. What you have is a 'heat pump', able to collect heat at a low temperature and discharge it at a higher temperature.

In the nineteenth century meat packers and other large-scale producers and shippers of perishable food used harvested ice in their business, but they also sought more controllable ways of cooling. The first commercially successful heat-pump refrigerator was developed by the Scottish-Australian entrepreneur James Harrison in the 1850s. It used fire in a steam engine with a flywheel said to be 5 meters in diameter to run the pump, and could produce three tonnes of ice in a day. In due course, however, electric motors supplanted steam engines to run compressor heat-pump refrigerators. By the 1930s they were small enough and quiet enough to use in domestic kitchens.

Even as Harrison was establishing a successful business with his design of heat pump, using a mechanical compressor, a French inventor called Ferdinand Carré was developing a different design, of which eventually several variants emerged, using different refrigerants. It also works in a loop, but the fluid is a gas such as ammonia dissolved in water. As it passes into the refrigerator-side of the loop, the ammonia evaporates out of the water, soaking up heat as it does so. It is then reabsorbed into the water, lowering the pressure in the loop. The water-ammonia solution flows into a section of loop outside the refrigerator, where it is warmed by a heat source, once again evaporating the ammonia. The warm fluid then discharges its heat into the room, and the ammonia redissolves in the water, ready to go around again.

The heat source which drives the cycle may be, for instance, a gas flame or other fire. On the other hand it may be, for instance, solar heat, rather than fire. An absorption heat-pump refrigerator is less effective than a compressor design; but it will operate

50

where no electricity is available. Moreover it can use any form of heat, including waste heat that would otherwise go unused. Using solar heat to provide cooling is a particularly appealing concept.

Managing information: with electricity

We humans communicate most directly with sound. Your vocal chords make the air vibrate all the way to my ear. But in order to speak and be heard, you have to be within 'earshot' of me. The breakthrough of electric telegraphy, creating long-distance communication that was almost instantaneous, spurred inventors to seek ways to transmit not just coded texts but actual voices with electricity. They wanted not 'writing at a distance' - telegraphy - but 'voice at a distance' - telephony.

The most successful was an experimenter and scientist called Alexander Graham Bell, born in Scotland, later resident in both Canada and the US. As a young man Bell studied speech, hearing and sound, seeking ways to help the deaf, who included both his mother and his wife. Prompted by the work of the German scientist Hermann von Helmholtz, Bell began trying to transmit the vibrations of sound not through air but along a wire, with electricity. He was not the only one. Other inventors and entrepreneurs jockeyed for patents and priority. However, in 1876 Bell narrowly beat out another claimant to gain the first patent on what became the telephone. Like the telegraph, the telephone in its early days used electricity from batteries. Only after telephone networks became widespread did they use electricity generated with fire.

Thomas Edison, too, studied the vibrations of sound, and indeed contributed a key improvement to Bell's telephone, a more sensitive 'microphone' to convert sound vibrations into a modulated electric current. But Edison also researched a different aspect of information and communication - the preservation or 'recording' of sound. Despite Edison's pioneering of electric light and electric motive power, his devices for recording sound used no electricity at all. Instead they captured sound vibrations mechanically, initially by means of a stylus inscribing a line on tinfoil wrapped around a cylinder turned by clockwork that you wound up by hand. This primitive arrangement could be played back by another stylus tracing the line and recreating the vibrations. From 1878 onwards, Edison and others greatly improved the recording devices and the media on which they recorded. But they were still hand-operated. Sound recording and playback with electricity did not arrive until the late 1920s. By that time, nevertheless, a new dimension of managing information with electricity was well established - entertainment at a distance, both in space and time. It would grow beyond imagining.

Sound moves through the air in waves. Scientists studying the behaviour of electricity and magnetism, in particular James Clerk Maxwell in the UK, suggested that 'electromagnetic waves' might also exist, moving not through the medium of air but through an imperceptible medium called the 'ether'. In 1887, a classic experiment disproved the existence of ether. Nevertheless, in 1888 the German scientist Heinrich Hertz demonstrated that he could indeed create such electromagnetic waves from a tuned spark gap, and send them from this 'transmitter' to a 'receiver', which exhibited equivalent sparks. With subtly elegant experiments he measured the speed of the electromagnetic waves, which proved to be the speed of light, confirming the suggestion that light itself is an electromagnetic wave.

Hertz died at 36, believing that his discovery would have no practical use. A young Italian aristocrat named Guglielmo

Marconi proved him wrong. Reading of Hertz's work prompted Marconi, aged just 20, to begin his own experiments. He quickly duplicated and expanded Hertz's findings, and at once foresaw the practical possibility of 'wireless telegraphy' - sending a coded electromagnetic signal through space rather than along a wire. Within only five years Marconi had developed equipment that could send messages not only across single countries but across the Channel and from a ship more than sixty miles offshore. By 1903, after some controversy and skepticism, including a long-running patent dispute with Nicola Tesla, Marconi had demonstrated wireless telegraphy across the Atlantic ocean. From then on Marconi's business, and his network of transmitters and receivers, expanded rapidly. In 1912 wireless telegraphy between ships helped rescue survivors from the wreck of the *Titanic*.

Just as telegraphy prompted people to develop telephony, so wireless telegraphy spurred work on wireless voice transmission, or simply 'wireless', as yet another way to manage information with electricity. Many inventors contributed. By 1920 the word 'radio' had come into common use, from the Latin word 'radius', suggesting 'radiating', or spreading out in all directions. Electromagnetic waves became known as 'radio waves'. Experimenters realized they could transmit a radio signal not only to a single receiver but to many all at once - 'broadcasting'. In the 1920s and from then on, radio broadcasting, of entertainment, news, music and propaganda, enabled by electricity, became a major cultural phenomenon in many parts of the world.

For managing information, electricity grew ever more versatile and ever more essential. Throughout the twentieth century and since, after the electric telegraph and the electric telephone, inventors and entrepreneurs have found ever more ways to use electricity for information. Thus far they have also given us wireless or radio, tickertape, television, teletype, telex, facsimile or fax, computers, email, the internet, mobile phones, server farms, data centres, social networks and smartphones, and the innovation appears to be accelerating. In the past three decades, converting language, numbers, sound and light into digital electronic form has dramatically expanded the possibilities for transmitting, storing and analyzing data. We are now grappling frantically with the implications.

Electricity expanding

By the early twentieth century electricity as a way to do what we do was becoming a fact of everyday life, for more and more people and more and more activities. Electric light was brightening nighttime streets in many cities and towns, and illuminating interiors of public buildings, offices, factories and more and more homes. Electric motors were powering machines in many factories, although not yet often in homes. Electric trams and electric motor vehicles were moving passengers and goods in many urban centres. The electric telegraph and telephone were carrying messages between individuals, companies and governments, over ever-longer distances into ever more remote places, and the electric 'wireless' - radio - was preparing to join the discourse.

Making all the electricity required was becoming a major enterprise in its own right. Electricity systems initially established to supply electric light diversified eagerly into supplying electric motors, spreading the load on the system into the hours of daylight. Electric vehicles, telegraph, telephone and wireless still mainly depended on batteries. But you now needed to charge many of these batteries. For electricity entrepreneurs this was a welcome addition to the catalogue of uses, yet another way to use electricity from the system during the day. Electricity system operators began to refer to 'load-building': persuading more and more people to use more and more electricity, at every hour of the day all year round. That in turn provided a stimulus for the operators to add more and larger generators to the system, taking advantage of the economy of scale in which doubling the size did not double the cost.

For electricity generators using fire, one obvious drawback of the traditional steam engine was that the piston of the steam engine moved back and forth, while the alternator had to spin. In 1884 the British engineer Charles Parsons came up with a better way to use fire and steam to spin an alternator. He called it a 'turbine', from the Latin 'turbo' - 'I spin'. The turbine had a rotating shaft, like the alternator. On the turbine shaft were mounted a series of blades, resembling a succession of windmills of increasing diameter, inside a close-fitting housing. You fed pressurized steam from a boiler into the housing at the end with the smallest set of turbine blades. The steam expanded through the series of 'windmills', making them turn the shaft, which then turned the alternator. The arrangement made much better use of steam than the back-and-forth steam engine. Within a couple of decades the steam turbine had largely supplanted the steam engine as a way to generate electricity using fire. The steam turbine, like the steam engine and alternator, also offered economy of scale. Striving to reduce costs, electricity operators ordered ever-larger steam-turbine-alternator sets, as the number of electricity users and the variety of their uses steadily increased the load on the system.

Not all systems, however, used fire. Some rivers offered falling water that could turn a 'water turbine', with blades akin to those of a steam turbine. If need be, you could dam the river, to create, in effect, an artificial waterfall. Hydroelectricity as demonstrated dramatically at Niagara Falls was soon powering lamps and motors and other electrical appliances in Scandinavia and elsewhere in Europe, as well as parts of Canada and the US with suitable waterways. The water turbine, too, offered economy of scale. Water turbines, dams and hydroelectricity systems, like those using fire, grew rapidly in size. In due course some electricity systems incorporated both types of generation, with fire and without.

Fire in the air

We humans have always envied birds. We can move on land and water, but not in the air - not by our own muscles, at any rate. Many legends, such as that of Daedalus and Icarus, record human attempts to fly. Leonardo da Vinci, among others, tried without success to devise a flying machine. The first successful attempt used fire. French brothers Joseph and Etienne Montgolfier began experimenting with balloons inflated with hot air from a fire in a basket underneath. In October 1783, after a series of trials with smaller balloons, first carrying ballast alone, then with a sheep, a duck and a rooster, the Montgolfiers built an enormous and spectacular balloon, ornate and elaborate, some 25 meters tall and 15 wide. In it Etienne became the first human to soar into the air.

However, even as the Montgolfiers and their supporters were making their early flights, another option was also emerging - one that did not use fire. In 1766 the British scientist Henry Cavendish first produced the gas hydrogen - what we now know as the lightest of all gases. French experimenters began developing hydrogen-filled balloons. In early December 1783, only a week after the Montgolfiers' hot-air balloon travelled nine kilometers over Paris, the Roberts brothers in a hydrogen balloon travelled 36 kilometers.

Both hot-air and hydrogen balloons nevertheless had a major drawback. In either kind of balloon you could soar high into the air. But you were then entirely at the mercy of the wind. You could not control what direction you moved over the land, or how fast. That seriously limited useful applications for balloons. Then came the 'dirigible' - which means 'directable' or 'steerable'. Like hot-air and hydrogen balloons, the dirigible could fly because it was 'lighter than air' - less dense, and therefore buoyant in air. But it also had motive power - one or more internal combustion engines turning propellers that would drive the dirigible horizontally through the air. The leading proponent of the dirigible was the German Count Ferdinand von Zeppelin, who became so identified with the concept that dirigibles became known generically as 'zeppelins'. Zeppelin created a design with a long rigid cylindrical framework of aluminium alloy, covered with an outer skin and containing a number of gas-tight bags that provided the buoyancy. From 1900 onward Zeppelin's large and cumbersome 'airships' suffered assorted mishaps, but attracted popular attention and enthusiasm.

At the same time, however, another approach to flight was rapidly emerging. After many inventors had tried and failed to fly with 'heavier-than-air' machines, in 1903 the American brothers Orville and Wilbur Wright flew their 'aircraft' a few meters high along a sandy beach in North Carolina. It was the beginning of rapid and hectic progress in 'heavier-than-air' flight. An 'airplane' used not gas-bags but wings - hence the 'plane' - to lift the craft into the air and keep it there. But it also relied on fire. For wings to work, the craft had to be moving horizontally, at a speed close to the maximum then achievable on land. The Wright brothers used fire in an internal combustion engine, turning a propeller to drive the machine. So did the other inventors who swiftly followed their successful example. Within a decade airplanes using fire were breaking one record after another, for height, distance and duration of human flight.

As airplanes grew ever larger and faster, their propulsion systems also improved. But the available internal combustion engines used pistons moving up and down to turn a propeller that had to rotate. The problem was similar to that affecting the steam engine driving an alternator for electricity, solved by Parsons with the steam turbine. The British inventor Frank Whittle found a similar solution for airplane engines. Initially intended as a better way to turn a propeller, it resembled a steam turbine, once again with a series of 'windmill' blades on a shaft. Instead of steam, however, this 'turbojet' used hot expanding gas from burning fuel to turn the windmills, the shaft, a compressor and a propeller attached to it. However, the design proved to have an even better function. The hot compressed exhaust gas, expelled from a nozzle at the rear of the engine at close to the speed of sound, made the engine behave like a rocket, thrusting the airplane through the air with no need for a propeller. The design was called the 'gas turbine', but we know it better as the 'jet engine'. It has since become still more versatile.

Electric diversity to electric monopoly

Organizing to supply and use electricity grew steadily more complicated. As the physical extent of systems steadily increased, so did the list of interested parties on each system. You might manufacture and supply the ever-larger hardware to generate, transmit and distribute electricity. You might purchase this hardware, either as a private entrepreneur investing in a business to sell electricity, or as a government, perhaps civic or municipal, investing in a facility to supply electricity to your constituents. You might produce and supply fuel to burn in power stations. You might own land on which others wanted to erect some kind of electricity supply installation - not only power stations themselves but also, perhaps, 'substations', nodes in the network, or transmission towers and lines, marching ever farther across the landscape.

You might design, manufacture and sell lamps, motors or other equipment to use electricity for our various human activities. Most important of all, you might want to pursue these activities - adjust temperatures, make light, exert force, move things or manage information - and you might want to do so with electricity. Whatever your interest, you now had to negotiate with many others to further it. Transactions to produce and use electricity became ever more complex. As a result, early electricity systems in different places emerged looking quite different one from another, in both technology and administration.

Take, for instance, the small harbour town of Hammerfest, near the North Cape of Norway inside the Arctic Circle. In 1889, in the dark Arctic winter, more than two months without sunrise, a Russian ship at anchor in Hammerfest harbour demonstrated its on-board arc-lights, astonishing the townsfolk. They were so taken with electric light that they sent two town officials all the way to the Paris exposition, where they ordered a water-turbine generator and all the requisite hardware to set up an electric lighting system for Hammerfest. Then, in the summer of 1890, before the system had been set up, a catastrophic fire swept through the town's wooden buildings. The Hammerfest town officials nevertheless made the electricity system their top priority, deciding that having electric light through the coming dark winter would greatly assist the task of rebuilding. They installed the water-turbine generator on the hillside above the harbour, channeling a cascade from the cliffs above. In early 1891 Hammerfest became the first European town to have public electric street lighting. The system was owned and operated by the town for its citizens. This arrangement soon became commonplace in many parts of Europe. If you wanted electric light or motive power you purchased your own lamps, motors and other equipment; but the town government supplied the electricity for you.

In London, by contrast, many different private electricity entrepreneurs competed for customers, stringing cables from building to building along London streets, often two or more companies seeking customers on the same street. They were proposing to run both electric lamps and electric motors; but different systems used different voltages and different frequencies, so that, if you used the wrong kind of lamp or motor, either it would not work at all or it would burn out, sometimes spectacularly. Across the whole of the UK electricity scene, confusion reigned for more than two decades, into the mid-1920s, until the UK government had to intervene, laying down standards and granting exclusive franchises to bring some coherence into the prevailing chaos.

In Chicago, Samuel Insull, the one-time private secretary of Thomas Edison, saw an opportunity. By the late 1890s he was running Chicago General Electric. A visit to London gave him the idea to charge different rates for electricity for different times of day, persuading customers to use more electricity at times when the system had unused capacity. He was thus able to bring down average bills, and win more and more customers. Local electricity systems to supply light and motive power were springing up in and around the city. By borrowing money at advantageous rates Insull was able to buy up these systems and interconnect them, making them steadily more effective.

He also convinced the local government that an electricity system was a 'natural monopoly': that having more than one set of wires in a neighbourhood was economically inefficient, since no single set of wires could then serve the maximum number of customers. The local government, thus convinced, gave Insull a monopoly franchise. No other supplier would be permitted to sell electricity in the franchise area. However, Insull also argued that such a monopoly franchise had to be regulated by the local government, lest it try to take unfair advantage of its captive customers by raising the price of the electricity it sold. This arrangement, a regulated monopoly franchise, soon became commonplace, not only in the US but wherever electricity systems sprang up.

After the revolution of 1917 that overthrew the Tsar of Russia, electricity became a central theme of the new Soviet Union under Vladimir Lenin. For Lenin, 'Communism is soviet power plus the electrification of the entire country'. Lenin was convinced that only under communism would such electrification, establishing electricity systems to supply users everywhere in the country, be possible.

In the US and elsewhere, however, programmes for rural electrification proved Lenin wrong. Throughout the 1930s the Rural Electrification Administration erected transmission lines and local networks over the vast areas of the US midwest. In Canada electricity systems owned and operated by the provincial governments did likewise. In the 1940s, in the remote Highlands of Scotland the North of Scotland Electricity Board built dams and strung transmission lines over the forbidding terrain to its scattered settlements.

In doing so, the rural electrification programmes overtook, and effectively eliminated, many local electricity systems. They substituted centralized supply, heavily subsidized by urban electricity users, for many systems on individual farms and other rural settlements, which had been based on their own wind and water turbines, under their own control. These local systems, however, could not compete with the heavily-subsidized supply from the central systems. The tension between centralized and decentralized - that is, local - electricity systems resurfaced half a century later.

War, fire, electricity

We humans have always killed each other. The Biblical story of Cain and Abel is typical, asserting that killing began in the first generation after Adam and Eve. Archaeological evidence of bones at burial sites shows many instances of violent death. Through most of human evolution, we have killed each other personally and individually. You met the person you killed.

Then, some 1200 years ago, the Chinese invented gunpowder. Initially for celebration, in fireworks and rockets, gunpowder gave a spectacular and harmless manifestation of fire. But a material such as gunpowder, an explosive, a solid that explodes in fire, expanding almost instantly many thousandfold, can expel a projectile with lethal velocity out of a gun - the 'gun' in gunpowder. A gun was a firearm, a weapon based on fire. Firearms transformed killing. Firearms, for instance, gave Europeans a deadly advantage as they moved into North and South America, Africa and parts of Asia. They could kill at a distance, never having to come into close quarters with their adversaries. With firearms, they did.

Successive designs of firearms made killing ever easier. With longer-range firearms you could kill someone farther away. With rapid-fire firearms, machine guns, you could kill many adversaries rapidly. To do so with fire, moreover, you did not need a firearm emitting single projectiles. You could use a bomb or a shell. In a bomb or a shell, when fire transforms a solid into a gas, the dramatic expansion creates a shock wave and hurls fragments of the bomb or shell casing in all directions, bringing destruction and death to anything and anyone nearby. You can also use fire to deliver the bomb or shell, by firing it from a cannon or mortar.

Throughout the twentieth century fire demonstrated its dark side with brutal frequency. In war after war firearms slaughtered millions. Fire from explosives in bombs and shells pitilessly destroyed cities and countryside alike, soldiers and civilians alike.

Electricity came late to warfare; but especially from World War II onward it too has played a role. Unlike fire, electricity in warfare is not lethal; but it increases the lethality of fire. Radar locates enemy aircraft and sonar locates enemy submarines, making them targets for fire. Radio communications aid troop movements and other logistics. More recently, electrical and electronic weapons-control systems have substantially worsened the destructive effects of fire, and made remote-control devastation feasible. Drones guided by electricity and powered by fire automate killing at a distance that can be thousands of kilometers.

Meanwhile an entirely new form of electric warfare is emerging, so called 'cyberwar'. It threatens disruption and even destruction through the internet and the vast array of computers, control devices and other electronics that now interact in real time worldwide.

Electricity regrouping

World War II wrought devastation, particularly across Europe and parts of Asia. Many countries that had already established electricity systems saw them shattered. But war also underlined the importance of electricity to modern societies, not only for convenience but for services becoming essential. For governments that already had electricity systems, rebuilding them became a high priority. Where electricity was still a rarity, creating electricity systems became a similarly high priority.

Ownership, operation and financial arrangements differed substantially from country to country. Almost immediately after World War II, for instance, the government of France created a state-owned company, Electricité de France or EdF, backed by French taxpayers, with total responsibility and authority for electricity supply countrywide. In Germany and Japan, however, no such measures were permitted. Both Germany and Japan, as defeated adversaries, were under occupation, particularly by the US. The US, as the foremost proponent of 'free enterprise', and deeply suspicious of anything that smacked of what US commentators called 'socialism', insisted that Germany and Japan had to entrust the rebuilding of their electricity systems to private industry - much of it, of course, from the US.

Arrangements in other parts of the world fell between these two extremes, with a varying and evolving mixture of government, regulation and private industry. Whatever the management regime, however, the physical shape of electricity systems worldwide rapidly converged to what became a common technical model, broadly the same wherever you looked. By the 1950s and onwards, electricity systems almost everywhere were based on ever larger turboalternators, steam-powered or water-powered, in ever larger generating stations in ever more remote locations. These huge stations produced so-called 'bulk electricity', sent out along high-voltage transmission lines that might be many hundreds of kilometers long, on tall towers dominating the landscape. The transmission lines in turn were connected to substations that reduced the voltage, and distributed the electricity by overhead wires or underground cables to users, especially in urban areas.

A lot of these generating stations used fire, burning coal. By the late 1950s systems were building stations a long way from users but right at the mouths of coalmines, calling the electricity they generated and delivered 'coal by wire', as though the fuel and the fire were more important than the electricity. Some stations used fire to burn the heavy oil left over from the refining that produced petrol, diesel and jet fuel, all now in increasing demand. Some stations, close to the oilfields producing petroleum, did the oilfields a service by burning the annoying and dangerous 'natural gas' that sometimes emerged with the oil.

Nuclear 'fire'

The discovery of nuclear fission in 1938 triggered hectic analysis of its implications. A 'chain reaction' in the rare metal uranium promised to yield a dramatic release of heat, much more extreme even than fire. If the chain reaction could be controlled it might therefore be a valuable alternative to fire. But initial concern focused on a more alarming possibility. A fast chain reaction, out of control, might produce an explosion far beyond any from fire. As the world toppled inexorably into yet another war, European scientists feared that Hitler's Nazi regime might win the race to create this terrible weapon, an 'atomic bomb'. The countries allied against Nazi Germany, led by the US, mounted an unparalled industrial undertaking, the 'Manhattan project'. In July 1945, in the New Mexico desert, the first nuclear explosion created what was to become its iconic symbol, the mushroom cloud. Less than a month later nuclear explosions over Hiroshima and Nagasaki brought a horrendous end to World War II.

Through the following decade several countries, including the US, the Soviet Union, the UK, France and China among others, devoted vast resources and effort to developing nuclear weapons.

That entailed mining and processing uranium, and building enormous and costly installations to do so. One spin-off from these programmes was the demonstration that you could indeed use a nuclear chain reaction to boil water and generate electricity. The Soviet Union did so in 1954 and the UK, with great ceremony, in 1956. The UK plant at Calder Hall, billed as the 'world's first nuclear power station', was actually built to make plutonium for UK nuclear weapons. But its 'nuclear reactor' did indeed also produce steam to generate electricity.

In the UK, France, the US and elsewhere, nuclear people were eager to promote civil nuclear power, not least as a way to compensate for the terrible threat they had created. But electricity companies were wary. They had cheap coal and heavy oil, fuels now familiar and abundant. They did not trust enthusiastic pronouncements such as that of the American Lewis Strauss, that nuclear power would be 'too cheap to meter'. But the nuclear people were politically influential, in part because of the weapons connection. From the late 1950s onwards they persuaded governments to launch nuclear power programmes in a growing number of countries worldwide. By the early 1970s some predictions of future expansion saw nuclear power superseding every other form of electricity generation, perhaps even by the end of the twentieth century.

It did not turn out like that. Nuclear fission appeared for a time to offer great promise as an alternative to fire as a source of heat, especially to generate electricity. But fission is a process more violent and extreme even than fire, with consequences that have proved to be acutely difficult and expensive to manage. Extracting the essential uranium fuel from ore has left many millions of tonnes of 'tailings', mountains of solid waste with radioactive contaminants such as radium that have poisoned waterways wherever uranium is mined, with no clean-up feasible. Radioactive spent fuel continues to accumulate in nuclear station cooling ponds around the world, while governments battle local authorities over disposal plans unresolved for many decades. Some nuclear power stations have proved economic, and performed well. But many have not, so much so that private investors now refuse to finance new nuclear plants without guaranteed and open-ended support from taxpayers and electricity users.

Insurers have always refused to offer coverage for possible nuclear accidents - understandable in light of the continuing, ugly and dauntingly expensive catastrophe at the Fukushima nuclear plant in Japan. Governments have stepped in, with legislation and treaties that free nuclear power from paying damages, leaving the bill with taxpayers. The first generation of nuclear plants is reaching, or has already passed, its anticipated service life. The countries now building new plants have governments with central control, such as China, or governments whose interest in nuclear technology appears to include not only electricity but also weapons. Fission still has powerful adherents and promoters, and a relentless drumbeat of publicity hails a 'nuclear renaissance'. But the reality on the ground is that costs continue to escalate, schedules continue to overrun, safety issues continue to mount and the capacity of plants shutting down continues to exceed the capacity starting up. After just over half a century fission's worldwide role in boiling water and raising steam to generate electricity is already declining. It is unlikely to recover.

Fire, electricity, environment

As electricity expanded into more and more human activities, we began at long last to notice the dark side of fire. In December

1952 the smoke and sulphur fumes from open fireplaces combined with London's familiar fog into a suffocating and impenetrable blanket of smog lasting four days, killing some 4000 to 12 000 people. Four years later the UK Parliament passed the Clean Air Act, banning open fires in urban areas all over the country, the first legal constraint on the use of fire. Instead of burning coal in urban fireplaces, huge power stations began burning it right at the exit from the mine, sending electricity as 'coal by wire' into towns and cities. Tall smokestacks sent the smoke and sulphur fumes high into the air, where they were expected to disperse harmlessly. As we found out later, they did not.

When making light, exerting forces and moving things, people began paying closer attention to what happened to their surroundings, the impact of human activities on the environment. Fire and electricity both caused environmental problems. Using fire to make electricity compounded the problems. Expanding electricity systems with power stations in remote areas sent tall towers or 'pylons' striding across the landscape. Many people, however keen to use electricity, said 'not in my backyard', or even on the horizon. Dams to produce hydroelectricity flooded scenic localities and disrupted river ecosystems. Power stations on lakes and rivers used their waters for cooling, raising water temperatures - so-called 'thermal pollution'.

Burning coal and heavy oil in power stations discharged smoke, fine particles, sulphur, mercury and other air pollutants. Cars and trucks burning petrol and diesel filled city air with carbon monoxide, nitrogen oxides, particulates and carcinogens. Producing and transporting oil to feed fires in boilers and internal combustion engines led to accidents with tankers and offshore rigs, causing oil spills that sometimes covered many square kilometers of the sea, fouling water birds and fishing grounds. Sulphur plumes from tall smokestacks drifted many hundreds of kilometers before falling as 'acid rain', poisoning waterways and killing forests, often in entirely different countries. Legislators, regulators and diplomats struggled to find ways to mitigate these increasingly serious consequences of using fire and electricity. It was an uphill task.

In 1988 an effect first identified almost a century earlier, that had been worrying scientists for decades, at last caught the attention of politicians. Fire releases carbon dioxide into the atmosphere, forming a thickening reflective blanket over us, gradually warming the surface of the land and the oceans, with steadily more alarming results. At a global conference in Rio de Janeiro in 1992 and many other conferences since, most of the governments of the world have sought to agree plans to try to get the consequences of fire under control. Thus far, however, implementing global measures has been impossible. Many poorer countries fear that environmental constraints will harm their economy. Powerful interests in some rich countries, particularly the US, refuse to accept the overwhelming scientific consensus. Fire continues to spread, its potentially catastrophic consequences ever harder to ignore.

What are we to do?

What we do is not the problem. The problem is how we now do it. We still do what we have always done. Much is what our Neanderthal forerunners did before us. We control heat flow, for survival and comfort, and for our countless other activities that now depend on it. We adjust local temperatures up or down, heating or cooling our food and drink, our materials and our surroundings. We make light, in ever more ingenious ways, to hold back darkness and lengthen our days. We exert force and move things, continuously, everywhere, at every scale from minute to gigantic, from microswitches to supertankers. Above all, as a new century flashes past us, we manage, we struggle to manage, an ever-expanding torrent of information, tracking, recording and analyzing everything we do, for good or ill.

To do all this we still use physical things - buildings, lamps, motors, electronics, fittings, appliances, vehicles - and the two processes - fire and electricity. We combine them in systems, interacting to deliver the activities we want. But we have let the parts of the systems drift grotesquely out of balance. Buildings that should keep us comfortable just by their design and materials instead require continuous injections from fire and electricity to compensate for inadequate structures. We are at last beginning to move on from Edison's incandescent lamp of 1880, which turned less than five per cent of its electricity into light; yet even now some defend that feeble performance as preferable. We exert force with electric motors, yet fail to incorporate the controls that maximize their effectiveness. To move things, on land, on the sea and in the air, we rely almost completely on fire, in petrol engines, diesel engines and jet engines, pouring their toxic exhausts into the air we breathe. Worst of all, we still make most of our electricity with fire, even though we need not, and now can have little doubt what fire is doing to us.

How has this come about? How can we rectify it, while we still have liveable cities and a habitable planet?

Part 2

The Trouble With Fire

How fire started

You are scared. With the rest of your tribe you cower in your cave, as the sky outside shatters. Brilliant flashes dazzle you. The air explodes, hurting your ears. Suddenly, not far from the mouth of the cave, a dead tree erupts in vivid light. The orange light flickers and dances, spreading from branch to branch. You watch, in dread and fascination. As the tumult overhead subsides, you venture warily out of the cave, toward the dancing light from the dead tree. As you approach, you feel a strange warmth, coming from the dancing light. To your surprise it actually feels pleasant, especially after the ferocious overhead onslaught now fading away.

Fortunately, you are watching the dead tree very carefully. You know how many dangers are out there, threatening your life and the lives of your tribe, and this looks like one of them. Without warning one of the high branches of the tree cracks and plummets down. You leap back as it hits the ground, sending a shower of bright sparks in all directions. Motionless, you watch the branch warily, as the dancing light spreads along it. Then you notice that the broken branch has fallen onto another dead branch, already lying on the ground. Gradually, as you watch, the dancing light moves onto the other branch, until it too is flickering brightly.

You are scared and wary, but you are also curious. You and your tribe have always been curious. Curiosity has caused you a lot of trouble. But it has also helped you discover new things to eat, and new ways to get them. Very slowly you edge forward, toward the pleasant warmth. Picking up yet another chunk of dead branch, you hold it away from you until its other end is in the dancing light. You watch, fascinated, as the dancing light begins to envelope the end of your dead branch.

72

The sun has set. But you have not been plunged into the usual threatening darkness that keeps you all alert every night, for the menace of hungry predators that think you are food. Instead the warm dancing light sends flickering shadows into the trees and bushes around the cave. Suddenly, as you peer into the shadows, you see two glowing eyes, high off the ground, something big, something alarming. You swing around, preparing to run, forgetting the branch you are holding. As you swing the branch toward the glowing eyes, they abruptly vanish, as the something crashes off through the trees. Whatever it was, the dancing light of the branch scared it away. You stare at the branch, your curiosity sharper than ever. If it will scare off predators, this dancing light is valuable. Besides, you like the warmth, taking the edge off the chill of nightfall.

But the branch you are holding is slowly disappearing. The light seems to be eating it up. The dead tree, too, has partly disappeared, and its light is growing sparser and fainter. You look around, in the light from your branch, and see other dead branches nearby. Carefully you pick up another dead branch and hold it in the light from the first one. Gradually the light spreads onto the second branch.

Behind you, your mate and one of your offspring are watching you intently, as curious as you are. They see the idea taking shape in your mind. They move cautiously forward, examining the ground and picking up other dead branches. They hold the branches in the light of yours, watching as the light spreads, feeling the pleasing warmth. Then the youngster reaches out to touch the flickering light. With a cry he drops his branch and thrusts his fingers into his mouth. You realize at once that if you touch the light it hurts. This light is potent. It pushes back the darkness, offers pleasing warmth, drives away predators - but you had better be careful with it, because it can also hurt you.

Other members of your tribe are now clustering around you, examining the ground nearby for branches or anything else they can pick up, that might share the dancing light. One picks up a stone, and holds it gingerly and hopefully in the light from your branch. But the light does not spread to the stone. Another uproots a green sapling from the ground, holding it in the light. It spits and sputters, but the light does not engulf the sapling. The light is clearly choosy about what it will eat. It prefers dead branches. Fortunately the forest floor around you is littered with them. Soon five or six of you are all holding branches with dancing light. You can feel the additional warmth they give off, a pleasing and welcome sensation in the cool evening.

This is too good to lose. You realize that if you want to keep the dancing light alive you will have to keep feeding it. You put your branch gently on the ground at the mouth of the cave. In the light from other branches you begin gathering still more branches, carrying them back to the cave and placing them near your original branch, but out of its light. Other members of your tribe follow your example. Before long you have a sizeable pile of branches, ready to feed the light before it finishes eating your original branch.

It's late, and the youngsters are sleepy. So are you. But someone must watch over the light, to feed it and keep it alive. You feed some more branches to the light. Then you seat yourself near it, leaning against the wall of the cave. It is going to be a long night. But it is warmer than usual near the light. Moreover, as you remember the eyes that ran away, you feel safer. In the morning you intend to gather more branches - a lot more branches. You can already see that feeding the light will be almost as demanding as feeding yourselves. But you want to keep it alive. Somehow you know it matters.

Fire, nature, humans

Was that the way that humans first acquired fire? We can never know. But the available evidence indicates that we probably got fire from lightning, possibly even before Neanderthals. Lightning, nature's raw electricity, still starts many thousands of fires every year. Volcano eruptions also start fires, but they are local, rare and so violent that humans would tend to flee rather than trying to preserve the fire. Lightning, by contrast, happens often, over much of the planet, in ways that may make fire available for nearby humans to gather up and nurture.

Fire has shaped our surroundings since long before we humans emerged. Charcoal layers in the fossil record tell of vast fires that incinerated wide areas. Localities in many parts of the world, including what are now Africa, North America and central Asia, evolved with fire as a frequent, almost regular event. A lightning-strike would ignite dry underbrush or grassland. The resulting fire would sweep through the area, burning everything that would burn, leaving a layer of ash rich in nutrients, clearing a pathway for the new growth that would soon spring up through the ash and regenerate the locality. The fire cycle might, for instance, be in step with seasons of dry and wet weather, fires following the drying, limited in extent by what was dry enough to burn, and dying out with the return of rain. Such fires became a part of the local ecosystem, cleansing and reinvigorating the soil and the terrain, as reliable and essential as monsoon rains.

When you first acquired fire you cherished it, keeping it alive, guarding it and feeding it assiduously, because if you lost it you might never get it again. You also watched as fire started by lightning opened clearings in the forest or swept across grasslands, leaving open spaces from which new plants soon emerged. In due course you realized that you yourself could start such fires. You could use your carefully tended fire to ignite, perhaps, the dense undergrowth around your cave, to burn it away, removing the cover that might otherwise allow dangerous predators to sneak up on you. You could open up patches of forest that might otherwise be impenetrable. You might even find that your fire had trapped, killed and roasted other animals, turning them into novel food dramatically easier to eat than raw meat - perhaps the origin of cooking. Sometimes, however, you started a fire that got away, raging beyond your control, destroying far more than you intended, perhaps even yourself and your family.

Fire wielded by humans became a way for us to change our surroundings, sometimes drastically, long before the dawn of recorded human history. We have been using fire ever since. The consequent changes are now growing ever more drastic.

Controlling fire

For Neanderthals and early *Homo sapiens*, the overriding concern about fire, once you had it, was to keep it burning. If you let it die you could not relight it. Starting a fire was, and is, difficult and challenging, if all you have is what you find about you. You have probably heard or read about the purported Boy Scout method of starting a fire, by rubbing two sticks together. If you have ever tried it you know it is futile. To start a fire you have to create a local high temperature, not just warm but too hot to touch, to ignite what you want to burn.

A key is to find something so light and flimsy you can heat it up quickly, say thin dry leaves or dry grass - so-called 'tinder'. As the Boy Scout method suggests, friction creates heat; but you have to concentrate it. One way is to take a hard pointed stick and spin it in a dent in softer wood, with the tinder packed around the point in the dent. You can spin the stick between your hands, but you may get tired before you get a flame. Alternatively, you can bang suitable rocks such as flint together to create tiny hot sparks, fragments of rock heated by the impact. If you can make a spark land in your tinder it may begin to smoulder and burn. Once again, however, you may be trying for a long, frustrating time before you succeed.

Once humans had iron and steel you could use what we now call 'flint-and-steel', banging a piece of flint with a piece of steel to get larger hotter sparks. In due course you carried flint-and-steel and tinder in a 'tinder box', a compact fire-starting kit in which the tinder was likely to be charred cloth, easy to re-ignite. The ancient Greeks used mirrors, lenses and glass vases filled with water, so-called 'burning glasses', to focus the rays of the sun to ignite tinder. Archimedes is even said to have used an array of mirrors to ignite and incinerate an attacking fleet of Roman ships - impressive but unlikely. Many philosophers through the centuries studied the properties of burning glasses. In the eighteenth century scientists including Antoine Lavoisier and Joseph Priestley used burning glasses in the research that led to the discovery of oxygen and its key role in fire. They established that fire is a process in which a material combines with oxygen, releasing heat and transforming the material chemically. A material able to sustain this process, a material that will burn, we call a fuel.

In 1669 the German alchemist Hennig Brand was trying to find the 'philosopher's stone' to transmute base metal into gold. By boiling urine, he became instead the first known discoverer of an element, the element phosphorus - 'light-bringer' from Greek, so called because in the presence of oxygen the white form glows faintly in the dark. It is also rapidly inflammable. By the 1830s phosphorus was being used in 'matches' called 'lucifers' - 'light-bringer' from Latin. You 'struck' a phosphorus match by rubbing it on a rough surface, once again using friction to create enough local heat to ignite the phosphorus. But white phosphorus is also toxic, and its vapour dangerous. Phosphorus matches caused so many industrial injuries, not to mention murders and suicides, that the red alternative form of phosphorus replaced the white, and 'safety matches' placed the red phosphorus on the striking surface rather than the matchhead, to eliminate accidental ignition.

When humans learned to produce and control electricity, you could at last replicate the service originally provided by lightning. You could create an electric spark, amply hot enough to ignite tinder and light a flame. At the time, however, in the early nineteenth century, no one apparently thought of starting a fire with an electric spark. People did not think of fire and electricity at the same time. They were different phenomena with different uses. Even when electric batteries became much more widely known, for instance for the electric telegraph, you would probably not find one anywhere near a fireplace, boiler or furnace. Nor could you ignite the commonest fuels, wood and coal, with a spark.

The first significant role for electric spark ignition came when engineers began experiments using town gas in the first internal combustion engines. You could make an electric spark jump across a gap between two metal points inside the cylinder of a gas engine. By doing so you could ignite a mixture of town gas and air inside the cylinder, and get not just a flame but an explosion. The complete combustion, almost instantaneous, triggered sudden expansion of hot gas, driving a piston to turn a crank and rotate a shaft. When petrol became available, engine designers once again resorted to electric spark ignition. It became a key feature of most designs of internal combustion engine. Rudolf Diesel, however,

78

showed that you could ignite one particular fraction of petroleum just by compressing it sharply, with no need for a spark or electricity.

Once burning, a fire itself keeps the temperature high, as long as you keep feeding it, replacing what it burns. However, as Neanderthals undoubtedly discovered and we know all too well, you may not have to feed a fire. Once started, it may feed itself. A local high temperature ignites anything nearby that will burn. A fire can spread with dismaying speed, consuming everything it can burn. For humans, Neanderthals and *Homo sapiens* alike, the most immediate problem with using fire once lit is keeping it under control. You can do that in three ways. You can keep everything that will burn well away from your fire; you can surround your fire with material that won't burn; or you can control the supply of air or oxygen with which the fuel reacts. Neanderthals probably relied on the first method, in their caves and shelters. We use the second and third methods, often in very complex ways, in our countless elaborate uses of fire. We mostly succeed, keeping fire under control; but when we fail the failure can be spectacular and devastating.

Fire around us

By the early twentieth century, at least in rich countries, starting a fire had become easy, an everyday routine in many different contexts. When you arose in the morning you laid a fire in the fireplace, or lighted the stove or furnace, to warm up the house. On the cooker, burning wood or coal, you boiled a kettle for tea or coffee. You probably started the fire with a match, and you saw the resulting flames, a genuine fire burning in your house, in a place where it was confined and where you yourself had to feed it to keep it alive. In the early morning or after sunset you lighted the gas lamps, now much brighter since the invention of the Welsbach mantle in 1885, the heat from the flame making the metal oxides of the fragile mantle gleam bluewhite. If you were fortunate enough and wealthy enough to own a car, you might have to light a fire in its boiler and wait for the steam before you could drive it. Alternatively you might have to heave a fiercely stiff crank over and over, until the spark ignited the petrol vapour and the pistons began to turn the shaft. Among your domestic appliances, however, you probably used fire only for your iron, a flat block of metal that you heated on the stove before smoothing your linen and apparel. Otherwise you relied on muscles, either your own or those of domestic servants, to sweep the floors, do the laundry, wash the dishes and carry out other household chores.

A century later, in the early twenty-first century, in rich countries, the pattern is dramatically different. To heat the house, perhaps, you turn up the thermostat on the central heating. That, for instance, opens a valve and triggers an electric spark to ignite a flow of natural gas in a boiler, producing a flame inside a metal cabinet in a cupboard somewhere inconspicuous. You do not see the flame and you do not have to feed it, assuming you have paid your gas bill. The thermostat may instead turn on an electric heat pump, drawing warmth from below your lawn, using electricity produced tens or even hundeds of kilometers away, perhaps using fire - or perhaps not. Under your kettle you may see a gas flame. You may, however, have an electric kettle, that may or may not need a fire somewhere else. Your lamps are probably no longer traditional incandescents dating back to Edison. More likely they are so-called 'fluorescents' or 'light-emitting diodes', 'LEDs', using much less electricity for brighter light. You also have a whole catalogue of domestic appliances, some with motors, such as vacuum cleaners, food processors and power tools, some with electronics, such as televisions, computers and smartphones, all using electricity almost certainly produced somewhere else, with fire or without.

Unless you are in a small minority your car runs on fire, invisible but noisy, a rapid succession of small explosions turning petrol or diesel into carbon dioxide, carbon monoxide, nitrogen oxides, polycyclic hydrocarbons and - in the case of diesel - tiny particles of soot, all of which the car exhaust pumps into the air around you. If you are in that small minority your car runs on electricity, made somewhere else. It may or may not use fire, and you probably don't know either way. Even when you do use fire you usually don't see it; and most of the time you don't even know you are using it.

In poor countries, however, and especially in their rural areas, even in the twenty-first century fire is still very much an everyday experience, and not a pleasant one. If you are, for instance, an African woman in a rural village, your daily chores may involve walking many miles to find dry wood to feed the fire on which you cook your family's food. The fire is inside your home, perhaps three stones under your cooking pot, with your scavenged twigs burning between them. The smoke from the fire fills the air around you and your youngsters as they play at your feet. As you prepare your food you all breathe the choking fumes. They will kill millions of you, women and children, every year.

Or perhaps you live in an Asian village. In the evening you have light, from a kerosene lamp that smells and smokes. The kerosene, delivered hundreds of kilometers over perfunctory roads, costs a significant fraction of your meagre income. But kerosene light is better than the darkness which is your only alternative. You inhale the choking air and pretend it doesn't bother you.

We humans have evolved with fire. Even before *Homo sapiens*, fire gave our Neanderthal precursors a power not available to other animals. Humans are the only animals that can start a fire.

81

That power has shaped human society, giving us materials and potencies humans alone can use. Fire is long since a fact of daily life, even now when in rich countries you see may open flame only on a gas cooker, a bonfire or a cigarette lighter. However, we have historically taken for granted, at least until the last half-century, the consequences of fire, in particular the smoke, particulates and gases it releases into the atmosphere. In rich countries we ignore them, or try to. In poor countries they kill.

Feeding fire

Ever since we acquired fire we have taken for granted the need to feed it, with fuel which it will rapidly consume. Feeding a fire is akin to feeding a human - perhaps even more continuously demanding. We have accordingly created social arrangements to support feeding fire, arrangements that far predate human history. Ever since humans acquired fire, gathering fuel to feed it has been part of daily existence almost as important as gathering food. Humans and fire both rapidly use up what feeds them. You have to replenish both food and fuel, almost continuously.

As society evolved and stratified, gathering fuel became a distinct category of labour. You might be a specialist, a woodcutter or a charcoal burner, producing fuel as a trade and a business. The concept of money emerged, to facilitate your transactions. In due course, in some parts of the world, society established the rule of law, laying down guidelines for public behaviour that no longer depended on the whims of the strongest. One typical family of laws allowed businesspeople to group together, to form companies with so-called 'limited liability'. Such companies could borrow money and make financial deals without putting the personal property of individual company members at risk. That greatly expanded the possibilities of business.

The business of some was to supply hardware, such as stoves, boilers and furnaces to raise local temperatures, lamps to make light, and steam engines to exert force and move things. The business of others was to supply fuel, such as firewood, coal and lamp oil. The companies selling hardware sold it as an investment. You the customer bought something durable, something that would last, for years or even decades. The companies selling fuel sold it as a commodity. You the customer bought something to use up, to consume continuously, feeding a fire. The distinction was, and is, profoundly important.

We were still doing the same things we had always done; but the arrangements were getting much more complicated. To make light, you no longer bought, say, a candle from one artisan and a flint-and-steel lighter from another, at prices you agreed with each seller. Now, instead of such simple transactions between an artisan and a customer for a price mutually agreed, making light might involve a customer, a company to supply the lamp, another to supply the lighter, another to supply the requisite fuel or electricity, and a government imposing laws and regulations with which all participants had to comply. Setting a price was no longer a straightforward bargain between seller and buyer. Moreover the costs involved were much harder to calculate, and you might not even know who bore them.

What was emerging was a system, a set of interlinked activities and interests that combined to enable us to make light and carry out our other activities. The different parts of this human-activity system, with different owners and interests, had to work together. We therefore needed agreements, transactions and compromises for the whole system to function, for us to do what we wanted to do. As time passed the system grew steadily more complicated. Companies to supply wood and coal to feed fires were local and small-scale, and often in direct competition with each other for customers, transaction by transaction. Then came gas companies, operating by means of contracts with customers, such that you got your gas from your neighbourhood company, and had little choice except to take it or leave it. Electricity companies. when they arrived, likewise operated by contracts with customers, although - at least in some places, such as London - you might have several companies laying cables in your street and competing to sign you up.

Whalers and their brokers on shore supplied whale oil for lamps. After the discovery of petroleum and its potential came companies supplying kerosene, swiftly supplanting whale oil wherever petroleum and its derivatives became available. That in turn entailed extracting and refining the petroleum, and transporting the crude oil to the refineries and the refined products to customers, not only for lamp oil but for motor fuel, lubricants and other products - an entire system in itself, including either railroads or pipelines or both. In the US, for instance, John D Rockefeller's Standard Oil expanded to embrace all the relevant companies and activities, becoming so powerful that the US Congress in 1911 passed a law to break it up. A century later its daughter companies, notably what is now ExxonMobil, remain among the largest in the world.

As our uses of fire grew more varied and more differentiated, feeding the various kinds of fire grew more demanding. In your steam engine, for instance, you could burn anything combustible that fitted into its firebox. But your internal combustion engine required a fuel meeting much tighter specifications, which depended in turn on how the engine ignited its fuel - an electric spark to explode petrol, sharp compression to explode diesel.

An intensifying dependence was developing, that has since become a profoundly important aspect of feeding fire. Once you have acquired, say, a particular engine, you can use it only if you also have continuing subsequent access to the particular fuel it demands. If that fuel becomes unavailable or too expensive, your engine, no matter how durable or reliable, is useless. It will no longer do what you want it to do.

The implications of this dependence, however, did not become serious until well into the twentieth century. Instead the oil business expanded rapidly, finding new oilfields in Texas and elsewhere in the US, across the Middle East and in due course in many other places. Petroleum and petroleum products proved surprisingly easy to transport and trade. The oil industry became spectacularly global. The result was an ever-increasing supply of ever-cheaper fuel to feed fires in internal combustion engines, factory furnaces, household heaters and power station boilers. The availability of this cheap fuel in turn elicited enthusiastic investment in the engines, furnaces, heaters and boilers that burned it. The system of hardware in which you fed fire with petroleum products expanded at a breathtaking pace, first in rich countries and then almost worldwide.

The combination of cheap abundant petroleum and internal combustion engines made the human activity of moving things, including people, steadily easier. Within a few decades we found, in many places, that we could move so easily and cheaply that we could live in one place, work in another, entertain ourselves somewhere else, and get our food, drink and manufactured goods from somewhere even farther away. We began to change the layout of society, separating residential areas from business areas from industrial areas from agricultural areas from recreational areas. Activities that once all took place within and around a compact village might now sprawl tens or hundreds of kilometers - much too far to walk or cycle with your own muscles. Your everyday routine now required a motor vehicle; and the motor required fuel to feed its fire.

Our age-old activities, and the things we did them with, became so spread out that we could no longer do without the fuel that originally promoted the spread. We became, in effect, captive customers for that fuel, especially petrol to feed fire to move things. On the other hand, we did not have to become captive customers for, say, household heating oil - except that cheap heating oil let us make cheap and flimsy buildings, which then likewise became dependent on a continuing supply of affordable heating oil, to feed the fires compensating for buildings unable to control heat flow.

That dependency, for oil products of every kind to feed fires of every kind, taken for granted and ignored, blew up in our faces in October 1973.

Enter 'energy'

The warning signs were there. The US played a central role in the creation and growth of the global oil industry. Five of the so-called 'Seven Sisters', the seven huge international oil companies that dominated the industry as it grew, were based in the US. In 1970, however, for the first time, the US became a net importer of petroleum. That gave new leverage to the Organization of Petroleum Exporting Countries, OPEC, established to represent the countries whose oil the international companies were extracting and processing. The countries wanted a greater share of the revenue their oil produced; but the companies refused to agree. Then, in October 1973, yet another war broke out between Israel and its Middle Eastern neighbours, including key members of OPEC. The US backed Israel. In response the OPEC countries, acting as a cartel, agreed to impose an embargo on shipments of oil to the US. Almost overnight OPEC succeeded in quadrupling the world price of petroleum.

The shock to the global system was traumatic. Around the world, if you relied on abundant cheap petroleum products to feed essential fires, for instance to run vehicles or heat buildings, your bills skyrocketed, if indeed you could find any petrol or heating oil to buy. In the US, drivers queuing for hours at petrol stations fought gun battles. Shortage of supplies of natural gas in parts of the US and industrial unrest in UK coal mines compounded the problem. Panic set in. Within a matter of weeks, listening to politicians and commentators describe what was happening, you heard a new narrative, focusing on what they began to call 'energy'. We faced an 'energy crisis'.

Scientists and engineers have used the word 'energy' since the English scholar Thomas Young first employed it in lectures to the Royal Society in 1807, as one of the key fundamental concepts describing our world and the universe around it. In late 1973, however, that was not at all what politicians and commentators had in mind. To them, 'energy' just meant all fuels plus electricity, a convenient shorthand.

In the first stage of panic, in 1973-74, governments and policymakers issued urgent exhortations to what they called 'energy conservation', by which they meant 'using less fuel and electricity'. In the UK the official slogan was 'SOS - Switch Off Something!' Before long people equated 'energy conservation' with freezing in the dark. Then, gradually, a different formulation entered the public discourse. Policy people began to refer not to energy conservation but to 'energy efficiency' - a much more upbeat and positive-sounding expression. Its practical implementation, too, was more constructive, entailing for instance improving insulation and using better appliances. Nevertheless its objective was still to reduce the use of 'energy' - that is, fuel and electricity. Energy efficiency told you how well something used fuel or electricity. But it said little if anything about how well the thing did what you wanted it to.

Calling all the different fuels, plus electricity, collectively 'energy' also entailed treating them as if they were all equivalent, as if one could directly replace another. After the OPEC 'oil shock' we had to find a 'substitute for oil'. Many declared that the answer would be to increase dramatically our reliance on nuclear power. In the US an official programme called 'Project Independence' called for rapid expansion of nuclear capacity. Not to be outdone, the European Commission proposed a fourteenfold increase in EU nuclear generation by 1985 - fourteen times as much nuclear power in just over a decade.

Few politicians seemed to realize the obvious inconsistency of these proposals. The most important and distinctive role of petroleum and its products was and still is to feed fire in transport, particularly motor vehicles. Nuclear power produces electricity. It was and still is essentially irrelevant for today's motor vehicles. Even for less specialized applications such as heating, you have to substitute not only the fuel oil but the entire system through which it flows, especially the hardware that delivers the heat. You cannot run an oil heater on electricity, or an electric heater on oil. Calling for nuclear electricity to 'substitute for oil' meant 'stop feeding fire' by resorting to an even more violent extreme process, which proved to be both dismayingly complicated and extravagantly costly. It did not happen. In the ensuing decade, however, it absorbed a staggering amount of money, skills and time, before gradually fading from the agenda.

The search for a 'substitute for oil' in the mid-1970s nevertheless set the pattern for future discussions of what politicians and commentators thenceforth called 'energy' and 'energy policy'. Using the word 'energy' as shorthand for all fire-fuels plus electricity allowed non-specialists, particularly politicians, to presume that they were all more or less the same commodity and interchangeable, that one could substitute for another, with no reference to the timescales or different hardware involved. It also meant that policy and commentary focused almost exclusively on fire-fuel and electricity, ignoring the rest of the system, the physical artefacts, the things that actually do what we want to do. That approach still prevails, distorting our priorities, limiting our options, crippling our decision-making and aggravating our vulnerability.

Electricity like fire

Smearing all the different fire-fuels together, adding electricity and calling the resulting melange 'energy' was novel. But we had been treating electricity as though it were fire for almost a century. Ever since the introduction of the electricity meter in 1885, electricity companies sold electricity by the metered unit, as though it were a commodity, like a fuel to feed a fire. The analogy was plausible. Edison's Pearl Street station near Wall Street, for instance, had a steam engine turning a dynamo, and the steam engine burned coal. Charging by the unit of electricity was equivalent to charging by the unit of coal burned to produce the electricity. That did not, however, account for the substantial additional cost of all the hardware involved - the steam engine and dynamo, cables, switches and of course the lamps that delivered the light, none of which were commodities being rapidly used up. Moreover, hydroelectricity such as that generated by the station at Niagara Falls did not involve a fuel at all, or anything to use up. The water was going over the falls anyway, whether or not you put a turbine in it.

Nevertheless, after the invention of the electricity meter, electricity systems evolved into a model closely akin to that for the gas-light form of firelight. Business arrangements treated electricity as a commodity, as though it were a fuel, to be consumed continuously. You bought and paid for it by the measured unit, just like a sack of coal. For most of the twentieth century, electricity settled into a pattern eventually established all over the world, a technical and commercial model like that of gas-light, that treated electricity almost as if it were fire. Indeed in many places electricity became a kind of fire at a distance. Some system operators generated electricity with ever larger steam turbine alternators, in ever larger power stations sited ever farther from users. Boilers burning coal, heavy oil or natural gas, or heated by nuclear fission of uranium, produced the steam to turn the turbines. Other operators dammed rivers, creating artificial waterfalls to turn water-turbine alternators. Both kinds of power station, with fire and without, produced electricity from a stored commodity, either fuel or an artificial reservoir of water, continuously using up the store as the station operated.

From each power station, long transmission lines operating at very high voltages carried the electricity, often hundreds, sometimes thousands of kilometers, to clusters of users in cities, towns and villages. The electricity, transformed to lower voltage, went through meters onto the premises of users. By this time, the tariff you paid usually included a fixed charge for the physical hardware leading up to the connection as well as a variable charge for the amount of electricity you used. The system was a monopoly franchise, the sole supplier of electricity in the locality. If you wanted to use electricity, you had to buy it from the monopoly. In many places, if you wanted instead to generate your own electricity, the monopoly would refuse to connect you to its system. It would declare that the refusal was for technical reasons, lest your generation interfere with the system. But the effect was always to make local on-site generation difficult, to reinforce the monopoly.

To keep the monopoly from otherwise abusing its privilege, for instance by overcharging, either the relevant government or a regulator it appointed decreed the tariffs. Ultimately, however, you as a captive customer bore the risk of whatever investment the monopoly undertook. In the 1960s and 1970s that risk became considerable. Striving for economies of scale with ever larger turbine-alternators, operators overreached themselves. Power station construction times overran schedules, sometimes by years. Eventual costs sometimes doubled or even tripled original estimates. A number of the largest stations in several countries proved not to work even close to their design specifications. But the company planners and managers responsible for the failures usually escaped the financial consequences, which fell instead on their customers.

Nevertheless, even as such problems began to accumulate, entrepreneurs and governments over much of the world replicated the central-station monopoly model of electricity, emulating the fire of gas light, with varying degrees of success, until the late 1980s.

Electricity versus fire

In the late 1980s, however, something striking began to happen to electricity. Even as the carbon dioxide problem was rising up the political agenda, the traditional model of electricity, with its resemblance to gas light and fire, began to break up. Ownership and control of the systems by which we use fire and electricity became a political issue, and intensely controversial. It has remained so ever since.

The initial impetus, particularly in the UK government of Margaret Thatcher, came from advocates of 'free markets', who deplored both the monopoly franchise and the role of government in electricity. As they assumed political power in several rich countries, they decreed that electricity systems owned by governments would be sold to private investors. They further decreed that the traditional centralized monopoly electricity suppliers would be broken up. The monopoly franchise would be abolished. Generators henceforth would compete for the right to sell their electricity to customers, just as though they were selling baked beans.

It did not work out quite like that. Abolishing the monopoly franchise meant that you the electricity user were no longer a captive customer. The risks of investment by system operators now fell not on you but on shareholders and bankers. Traditional coal-fired and nuclear power stations, of enormous size, might take upwards of six years to build and bring into service - risky investments, when no one could foresee with confidence how much electricity they could then sell, or at what price.

A new form of electricity generation was also about to arrive on the scene. The gas turbine or jet engine had always operated in short bursts, as required by, say, a long-haul flight between continents, or for brief periods of emergency use in power stations. By the 1980s, however, larger and sturdier gas turbines were proving that they could operate continuously, with no need for frequent maintenance, for many months at a time. Liberalization of electricity thus happened to coincide with the successful demonstration of the industrial gas turbine for continuous operation, and the arrival of abundant cheap natural gas to fuel it. Moreover, you could use the hot exhaust gas from a gas turbine to raise steam for a nearby steam turbine, in so-called 'combined cycles'.

Traditional electricity assumed that a better power station was always a bigger one farther away. With the gas turbine, however, a smaller, better power station, clean enough to locate close to users, could be built in two years or less. Almost overnight, anywhere you could obtain cheap natural gas, the preferred design for new power stations used gas turbines and steam turbines together, gas-turbine combined cycles or CCGT, sometimes even in urban areas utterly unsuited to any traditional power station.

This break with tradition changed the criteria by which you operated and enlarged the network. It also coincided with striking progress in generating electricity using not fire but wind, water and sunlight, in much smaller units but many more of them. Traditional steam and water power have always been based on anticipated economies of scale. But innovative smaller-scale generation exhibits economies of series manufacture, with a swift learning curve. Unlike the fire-based model of traditional generation, consuming fuel or stored water that you have to replenish continuously, innovative generation uses natural forces. The popular name for this electricity is 'renewable'. A more informative description is 'infrastructure electricity'. You invest in a physical asset such as a wind turbine, a microhydro turbine or a solar array. It becomes part of the infrastructure - infrastructure that generates electricity to use in any activity you desire. Harvesting natural forces it needs no fuel, nor does it consume anything. It does not fit the traditional model of electricity based on or emulating fire. You could also therefore call it fire-free electricity.

Decentralized, fire-free infrastructure electricity offers a dramatically different way to adjust local temperature, make light, exert force, move things and manage information. The central concept of this innovative system is not a commodity but a process. The transactions involved need not be batch sales of a commodity. They can be longer-term contractual relationships with matching financial transactions. Someone invests to set up the system, and someone else pays for access to the process when it is running.

Since a number of countries liberalized electricity more than two decades ago, decentralized generation has become a significant and rapidly growing contribution to total commercial electricity supply worldwide. But much of it sits uneasily with traditional centralized fire-model electricity, for several reasons. To the traditional system, an offshore windfarm looks much like a conventional central-station generator. But photovoltaic panels or cladding on individual buildings, or other decentralized small-scale generation, does not. You connect it at low distribution voltage. It looks more like what might be called a negative application, putting electricity into the system rather than taking it out. In doing so, moreover, it can reverse the current flow in nearby circuits, confusing protective devices and otherwise complicating system operation.

Yet more troublesome is the question of finances. What value does decentralized generation contribute? Who benefits, who pays, on what basis and how much? These questions are becoming steadily more pressing as the role of decentralized generation steadily increases. As the technical model of electricity systems evolves ever more rapidly away from the traditional fire-based model, institutional and business arrangements are becoming rapidly more incoherent and unstable.

Firefight

As the stress on traditional fire-based electricity intensifies, it is aggravated by a fierce counterattack in defence of fire. That is not how it appears, but that is what it means. Many politicians and commentators, for instance, unmoved by the overwhelming scientific consensus, insist vociferously that what is happening to the world's weather systems has nothing to do with our human activities, nothing to do with the way we use fire. Despite the mounting evidence to the contrary, they have convinced many of our fellow humans that we need not be concerned, nor make any change in our activities. Many analysts, including official national and international agencies, argue that we cannot do without ever-increasing use of fire, and the fuel to feed it. Again and again they produce projections that anticipate continuing spread of fire across the globe, regardless of the inevitable consequences.

The battle grows steadily fiercer. Confronted by ever more dismaying evidence of both local and global damage from the consequences of fire, many governments around the world have - in effect - declared their intention to eliminate fire from most human activities within the coming half century if not sooner. Electricity is a particular focus of their plans, some calling for a complete elimination of fire-based electricity within three or four decades. But feeding fire is one of the world's largest economic sectors. Huge companies, not to mention entire countries, draw much of their revenue from producing and selling fuel to feed fire. The companies sometimes assert that they too are concerned about fire effects. But they then behave as though they do not believe government declarations about future limits on fire and fuel. They continue to invest in new production facilities for coal, oil and natural gas, and in power stations to burn them, facilities expected to remain in service for decades to come. They claim that the fuel resources they own are worth hundreds of billions of dollars. They expect in due course to sell them to be burned.

The companies are understandably unhappy about suggestions that they are poisoning cities, acidifying oceans and threatening the planet. They insist, with some justification, that they are doing what society wants them to do. If society wants them to change, society itself must change, and change the ground-rules that still make fire central to human activities, despite its destructiveness.

How can we change the ground-rules? How can we get fire under control, before we let it destroy the human civilization we have created? How can we move beyond the Fire Age?

Part 3

How We Can Do Better

Better then

You are curious. You are always curious. You watch and listen, and sometimes you think about what you see and hear. This time you are watching the birds darting into and out of your cave. You have seen them outside, swooping into the puddle left by yesterday's rain, scooping up mud in their beaks, and flying up to the rock overhang above your head, just inside the cave entrance. You can't actually see how they do it. It happens too fast. But you can see what they are doing. They are adding blobs of mud, one after another, to a small round shape of mud about the size of your cupped hands, stuck to the rock wall. You know what it's for. You have seen them do this before, soon after the snow melts, again and again. Soon you will hear the familiar chittering noise of baby birds inside the cup of mud. If you keep watching, before long you will see the babies fly away.

You are curious, because you have just had an idea. If those little birds can make something out of mud, so can you. They only have beaks. You look at your two hands, each with fingers and a thumb. You can pick things up and hold them. You can certainly pick up mud. What if you picked up mud and put it on the ground just at the mouth of the cave? If you had enough mud you could do what the birds do. You could pile up handfuls of mud to close up most of the hole into the cave, leaving just enough room for you and the other members of your tribe to get in and out. That would keep the wind from blowing so strongly into the cave, would make the cave more comfortable, especially for sleeping. A mud wall might also help to fend off unwelcome nighttime prowlers, especially if your precious fire were directly inside the remaining opening. A mud wall would keep in more of the warmth from the fire, too - although it would also keep in more of the smoke, not such a good effect.

Squatting in the entrance to your cave you look from your hands to the mud-puddle to the side of the cave, and shake your head doubtfully. It won't work. The puddle is too small - not enough mud to be much use, not for you at any rate. The birds are doing fine, but they are tiny. For a better shelter than the cave you now have, you and your fellow Neanderthals will need much more material, enough to enclose a space big enough for all of you, infants, adults and elders, your sleeping places and your fire, a more comfortable, more convenient space. Mud alone will not suffice.

Still, the birds have made you think. You now realize that you can imagine being more comfortable and maybe even safer in a different shelter, maybe even one that you yourself have made for the purpose, not just a hole in a hillside that your elders found long ago. From now on you will be paying closer attention to what you have around you, thinking of possibilities, thinking of making things - better things.

Was that how it started? We shall never know, but it might have been. Neanderthal or *Homo sapiens*, somehow, somewhere, you hit on the idea of making things - clothing probably first, then shelter. As an early human maker you are emulating the other living creatures with whom you share your surroundings. They too make shelters - nests, dens, burrows and so on - with the materials they find around them. They do so to provide better comfort and better security for themselves and their families. So do you. What they are doing, and what you are thinking of doing, above all, is controlling heat flow, making things to control it better, to keep yourselves warm.

With mud, branches, skins and whatever else comes to hand, you fabricate an enclosure of some kind, perhaps walls and roof, strong enough to support itself, solid enough to keep out the cold and keep in some warmth, and with enough room inside for you and yours. In subsequent years, centuries and millennia your descendants will discover, invent and design endless improvements. Controlling heat flow with the structures we make will become so routine we take it for granted.

You also learn to add warmth, to raise the temperature around you, to make you more comfortable. Just having more members of the tribe inside your shelter helps, each contributing body heat. Once you begin domesticating animals, their body heat, too, can contribute. Dogs are particularly manageable indoors. If you get too warm you just move the dogs outside. But you have no way to lower the temperature further. You also have no way of knowing that doing so might, for instance, keep your mammoth meat from rotting so fast.

To cook the mammoth meat you use your carefully guarded fire. However, to raise the temperature inside your shelter, the fire is more problematic. It certainly works. But the heat it releases is dangerously hot. If you get too close it will burn you. Inside a shelter, moreover, fire produces smoke and fumes that sting your eyes and catch in your throat. Nevertheless, despite its drawbacks, especially indoors, fire is the only way that you can make light. Many millennia will pass before you find a better way.

The only way you can exert force and move things is with muscles, your own and those of other humans. In due course, when you succeed in domesticating animals, you will be able to use their muscles to pull, push and lift, for the forces you want to exert. Simple machines, things you make - the lever, the wheel-and-axle, the ramp or inclined plane - will multiply and redirect the forces of muscles, making them work better. Some humans will also coerce others, compelling them to serve as slaves - better for the slave-drivers, not so for the slaves. Eventually, only three centuries ago, you will at last devise and construct a machine that can use fire to exert force - the steam engine. The consequences will transform human society, for good and ill.

You and your fellows communicate and share information with voice and gesture. Once again, many millennia will pass before you find better ways, to transmit information beyond line of sight and earshot, and to record and preserve it for future use. When you do, you will do it with things - brush and paint, hammer and chisel, pen and parchment, then ever more elaborate and complex devices and structures.

Over the millennia, making things, making things for human activities and then making better things, will gradually but steadily widen the chasm between us humans and all other animals.

Better now

We humans make things, but not all of us are good at it. Almost all the things I use in my daily activities have been made by other people, almost all unknown to me and probably half a world away. I am nevertheless among the fortunate minority, much luckier than most of my fellow humans. I have clothes for most circumstances and most occasions. I can choose what to wear, to keep me warm and comfortable no matter what the weather or the ambient temperature. I spend most of my time inside buildings whose design and construction I had nothing to do with. My own house I can operate. I can adjust temperature and other conditions to suit me and my family, using doors, windows and other features, as well as fire and electricity. In other buildings, however, I have to accept whatever temperature and conditions the operators elect to maintain, although I may perhaps grumpily don or doff clothing if the temperature is too low or high. The same applies to lighting. My own lamps, on my own premises, I can operate, buying and paying for the electricity, generated

101

somewhere else by someone else. Otherwise, both indoors and out, I have to accept the ambient illumination, some combination of sunlight or moonlight plus whatever lamps other people operate.

Most of the time, when I am exerting force, I am doing so with my own muscles, to move things, first of all myself. When I was younger and more fit I cycled everywhere. Now, however, I mostly move both myself and other things with fire, in the engine of my car and in those of the buses, trains, aircraft and occasional ships I travel in. Because I'm not good at making things I only infrequently use tools to exert force, except for instance a tin opener, a bottle opener or a carving knife - none of which requires fire or electricity to operate. But I do have both an electric drill and an electric chainsaw, which can exert considerable force - indeed, in the case of the chainsaw quite alarming force, so much so that I was long reluctant to buy it. The commonest force I exert on any other object is probably the gentle force I exert on the keyboard and mouse of my computer.

That demonstrates the human activity now becoming ubiquitous, at least in rich countries - managing information with devices using electricity. Like so many of my fellow humans I now spend hours each day on my computer or my smartphone, working, communicating or playing, relying not only on the devices themselves but also on their batteries and the electricity supply that charges them.

All these activities, everything I do, I do more or less routinely, regarding them as a matter of fact. So do most people fortunate enough to live in circumstances akin to mine. We take these activities for granted, what we do, how we do them and with what things. But we could do so much better. I have been thinking, talking and writing about this all my professional life, for more than four decades. On the one hand I am ever more dismayed at the mounting evidence of the harm our everyday activities are doing to ourselves and our only planet, especially by our reckless use of fire. On the other hand I am puzzled and frustrated when I see the abundant opportunities we are failing to seize.

In my own activities I try for improvements. I've done so for decades. In the mid-1980s, for instance, we switched from incandescent lamps to the then-new compact fluorescent lamps - CFLs. In our main sitting room the conversion reduced our electricity use from 460 watts to 61 watts. Now we're switching to light-emitting diodes - LEDs. In the main ground-floor corridor, with the new LED lamps, the electricity use is down from 120 watts to 16.8 watts.

We've insulated lofts, ceilings and cavity walls, replaced single-glazed doors with double-glazed and built porches at two entrances. Insulating the cavity walls meant we could turn the thermostat on the central heating down 2 degrees Celsius, 4 degrees Fahrenheit, and remain comfortable even mid-winter. But our gas-fired boiler is old, and not condensing; a lot of the heat from the fire escapes out the flue in water vapour. However, the boiler still works; and I've yet to bring myself to junk it and install a better one. The hassle factor and the need for significant investment keep me from making an upgrade I'm sure would be advantageous. Despite what many economists claim to believe, we humans do not always act rationally.

In any case, my own choices can have at best a vanishingly small impact on human-activity systems in general. I could fly less, drive less, eat less meat - all the well-known varieties of self-denial offered as ways to reduce my impact on the planet. But I see no likelihood that enough other humans will opt for self-denial as a way for us all to do better. In any case far too many humans, including far too many in rich countries, are already denied far too much, denied adequate or even minimal opportunities to make their daily activities less arduous and more rewarding. We need a way that works for the many, not just the few.

Moreover, the choices available to you and me personally, while significant, are also constrained. We are caught up in systems with complex interconnections, using obscure resources and relying on elaborate supply chains about which most of us know little. We are also the focus of aggressive business models and the target of relentless advertising, trying to shape our expectations and decisions, while taxes and other financial frameworks jostle us.

I am bombarded almost continuously with offers from others eager to sell me the things they have made. They usually insist that what they are selling is better than what I already have. I am usually unconvinced. What I think is 'better' may not be what the advertisers and marketers tell me is better. The manufacturers employ the advertisers and marketers. The manufacturers design and make things they want me to buy and use; but their criteria are often not mine. The things they want to sell me are all too often designed to be cheaply made, short-lived and impossible to repair or maintain - in short, to be soon discarded and replaced through yet another purchase, over and over. That's not what I call 'better'.

Nevertheless, even against this depressing background, developments are encouraging. They persuade me that we are steadily doing better, but that we need to go faster. The way ahead is straightforward, a way to do better according to the criteria that seem to me the most important. We have to make our human activities easier, safer, cleaner, fairer, more reliable, more rewarding and less expensive, taking into account all lifetime costs, in money, resources, skills, time and consequences. Above all we must improve our physical things, especially the things that do what we want to do. We must reduce and eventually minimize waste of all kinds, with better materials, better design, and better integration between things. The better the things, the less we need either fire or electricity. At the same time we need to shift from using fire to using electricity in our activities, and shift from fire-based electricity to electricity not based on fire - to fire-free infrastructure electricity.

How can we do so? How can we do so in time? If we survey what is already happening, we may see how to speed it up.

Controlling heat-flow better

How we do what we do depends on what we make to do it with - how we make the things we use. Start with the single most important human activity. Unfortunately it is also the most boring. When you are doing it properly it hardly qualifies as an activity. All you do is put the right things in place, and they control heat-flow for you. The things that do the job are probably not photogenic. They may not even be visible. When they are visible, perhaps in the form of clothing, what you see is their appearance, not their function. Indeed, the most visually striking forms of clothing tend not to keep you very warm. In the case of shelter, you can rarely tell just by appearance whether the materials used are very good at controlling heat-flow, even when you can see them. Brick and stone, for instance, look reassuringly solid. But they both conduct heat pretty well, imposing only a modest barrier to its flow. Brick or stone alone will not keep you very warm.

To control heat-flow better, then, the first step is to select the right materials. Clothing you probably select every day. You do so mostly by experience and instinct. You don't wear a swimsuit in the Arctic, or a parka in the Sahara. If you expect to be in extreme conditions, with temperature very low or very high, you wear clothing designed to cope with those conditions. In very cold surroundings you wear thick quilted fabric made of materials such as wool or kapok or other artificial fibres, with many tiny air pockets among them to impede heat flow. In very hot conditions, by contrast, you wear loose light fabric to allow air to circulate around your skin and carry away your body heat. The fabric should be white, for preference, to reflect away sunlight and radiant heat.

Most of the time, for most people in rich countries, the role of clothing to control heat-flow is routine, and fairly easy to optimize. It mostly goes unnoticed. Much more important, indeed crucial, is another aspect of controlling heat-flow, not personal but collective, in shelter - in buildings. Here again you control heat-flow with things. In buildings, however, most of the time, you are not the one who chooses. If you were, you could choose materials of the right type and thickness to impose an effective barrier blocking the flow of heat, to keep the interior comfortable whatever the temperature outdoors. However, far too many buildings of every kind now do this poorly. Those who do choose choose badly. Instead of relying on the materials and structure of the building to keep heat in or out, they resort to fire or electricity, to replace desired heat the building leaks out, or to run air conditioners to pump out unwanted heat the building leaks in. Why have we let this happen? Why have we built so many inadequate buildings all over the world, when we know better? What can we do about it?

In many places, and certainly in rich countries, when you propose to erect a building you have to meet certain minimum standards. The building has to stay up. It has to have adequate air circulation for those inside. It may have to include, for instance, pipework for fresh water and also for waste water. It may also have to meet a standard for heat-flow control, for insulation. The word 'insulation' comes from the Latin 'insula', meaning 'island'. The building should be a 'heat island', such that heat cannot easily flow in or out of it. That means using both the right materials and enough of them, with consequent costs. Some countries with cold winters, such as those of Scandinavia and central Canada, set high standards for building insulation. Other countries, such as the UK, are much less demanding. Standards, moreover, are only effective if enforced. Cutting corners cuts costs, at least superficially. Over the years evidence suggests that even the modest insulation standards laid down, for instance, in the UK are enforced only feebly. Elsewhere, especially outside the rich countries, the enforcement problem is yet more acute. Buildings leak heat both out and in.

We are gradually tightening both the standards themselves and their enforcement. Even so, when estimating the cost of a new building, we often add up what it will cost to build, but fail to include what it will cost to run. When, for instance, you buy a new car, you take account not only of the sticker price that you will pay the dealer, but also of how much it will cost you to run, particularly how much fuel it will burn per kilometer traveled. For some reason, however, we seem not to think this way about buildings. We hear a lot nowadays about what politicians and the media call 'affordable housing', by which they mean houses comparatively cheap to buy. Making a house cheap to buy, however, usually means using as little material as possible. That in turn means that the house will be incapable of keeping you warm in winter or cool in summer unless you use, and pay for, a lot of fire and electricity to compensate for the inadequacy of the house itself. Making a house affordable to buy may make it unaffordable to run.

Many of the buildings with the poorest heat-flow control have been built in recent decades. Some, often the largest, most eye-catching or obtrusive, are just architects showing off, with little or no pretence at heat-flow control. In others, architects and designers appear to have considered a trade-off between cost of construction and running cost, and decided that the savings on flimsy construction would outweigh the extra running cost. When the costs of fuel and electricity were reliably and consistently low, such a trade-off might have made sense. But the costs have proved to be both unpredictable and high. Moreover conventional cost estimates always omit the costs associated with using fire to compensate for inadequate materials - an omission increasingly indefensible. However, if you are the one who pays to build the house or other building, you may be expecting someone else to pay to run it. You're happy to keep the initial cost down, by skimping on materials, because you won't have to pay the fuel and electricity bills.

That has left us with billions of inadequate buildings around the world, far too many of them in rich countries where we know better and can afford better. If we are to control heat-flow properly, so that we no longer use so much fire to compensate, we have to get serious about upgrading existing buildings of every kind. Residences, offices, schools, hospitals, hotels, shops and shopping malls, railway and bus stations, airports, warehouses, leisure centres, industrial buildings, government buildings - every structure that has humans in it can do a better job of keeping them comfortable. Except in extreme conditions the structure alone can probably do it, given body heat from occupants and residual heat from lamps, motors, appliances and electronics, relying little if at all on fire.

But different types of building pose different problems, and offer different opportunities for improvement. In practice, because each individual building is unique, in location, scale, configuration, age, use, history and so on, you have to tackle each individual building on its own terms. You have to assess how it functions now, where and why it falls short of its potential, and what you can do in practical ways to make it better. You may then have to add to and change the physical structure, essentially a further phase of construction. That takes investment, skills and time. It also clashes with any activity for which you are already using the building. All these obstacles hamper and impede the progress we urgently need. Somehow we have to confront and overcome them.

In the case of residences, we now have a rapidly expanding catalogue of houses and other residential buildings specifically designed and built to control heat-flow with little or no use of fire. In Germany the concept is called the 'passive house', 'passive' meaning that you need no 'active' heating or ventilation, no active process, no fire nor electricity, to keep your residence comfortable whatever the season or the weather. Some early designs with this objective looked frankly bizarre, with odd shapes, tiny windows and forbidding exteriors. They made the concept of effective control of heat-flow by the building itself seem outlandish. Since those early days, however, you can now find many passive houses not only in Germany but elsewhere in Europe and North America, which pass unnoticed. They look exactly like the other houses nearby; but if you live in one your heating bills almost vanish. Experience suggests that the cost of more expensive materials, such as thick insulating walls and high-performance windows, is often largely offset by eliminating the need for active heating and ventilating equipment, even before the saving on running cost.

However, once a house has been completed and occupied, improving its performance is more difficult. How can we make existing conventional houses, so many billions of them worldwide, perform more like passive houses, controlling heat-flow without needing fire to compensate? Many studies have concluded that the procedure is usually technically straightforward. The commonest shortcomings are thin, poorly-fitting windows and doors, and insufficient insulation in walls and roofs. Both are usually technically fairly easy to rectify, if the structure is otherwise sound. But you have to want to do it, you have to have the money to invest, and you have to be willing to put up with at least some inconvenience while the work is done.

That means that the problem of controlling heat-flow better in houses is not so much technical as financial, institutional and psychological. Even the financial aspect is often mainly psychological. Most upgrades will soon save enough on heating bills to pay for themselves. But you do have to put up the money at the outset, and believe that the savings will accrue. Moreover, many people living in the flimsiest dwellings have little if any spare cash available beyond their daily living expenses. When faced, as too many are, with choosing between heating and eating, they do not have the option of investing, say, in insulation, because they simply do not have the money. That becomes an institutional and political issue. In the UK it is called 'fuel poverty', and it is a major challenge, in part because the UK's housing stock is among the poorest in Europe at controlling heat-flow.

The institutional problem we face is common to all human activities, but it is especially acute in our approach to controlling heat-flow. We should be doing it with things - with buildings designed and constructed well enough for the purpose. If they are not, we should do something about it. However, those who make the choices and decisions in our society more or less worldwide are still so preoccupied with fire and electricity, and especially with feeding fire, that they pay at best only incidental attention to the things that do what we want to do. That failure is especially egregious for buildings. Study after study, report after report, has demonstrated how much fuel and electricity we waste in buildings of every kind, how we can upgrade them to improve their performance, to make them more comfortable and congenial for their occupants, and how rapidly most building refits will repay their costs in lower bills for fuel and electricity. Yet we continue to do very little about it, certainly much less than we could.

The studies always identify assorted barriers that hamper attempts to upgrade buildings. Buildings differ over a vast range, from single-family dwellings to skyscrapers. Measures that make sense for one may make no sense for another. But dealing with individual buildings one by one needs specialists and time. The initial cost deters those with little cash to spare. The hassle factor, putting up with some inconvenience in your home while work is done, puts some people off. If you are the landlord, you generally don't pay the fuel and electricity bills, in which case you have little incentive to improve the building. On the other hand, if you are the tenant, you usually pay the fuel and electricity bills; but you will be reluctant to invest to improve a building you do not own, and indeed you may not be allowed to.

Some studies refer to what they call the 'rebound effect': if you save money on heating your home, you may choose to maintain a higher temperature indoors, or you may spend your savings on other activities that use as much or more fuel and electricity. If the aim of a building upgrade is, say, purely to minimize the use of fire, the rebound effect may therefore reduce the effectiveness of the upgrade. On the other hand, if you are the occupant, you may get the benefit of more comfort or other rewarding activity in return, making the upgrade worthwhile even with the so-called 'rebound'.

All these impediments do undoubtedly keep us from making much more rapid improvement in the way our buildings control heat-flow. But they can be dealt with. The single most stubborn impediment is much more fundamental. It is the worldwide official and commercial fixation on supplying fuel and electricity, what they call 'energy', regardless of how much we waste. Official and commercial policy for our human-activity systems is equivalent to opening the bath taps ever wider, without putting in the plug. That is the underlying problem not only for controlling heat-flow but for all six of our human activities. In everything we do, governments, institutions, business, academics and the media are preoccupied with maintaining the processes, fire and electricity, while paying only passing attention to the things, the physical things, we use to do what we want to do. To get fire back

111

under control, our first priority must be better things - above all better buildings.

Heating better

Most of the heating we use, especially in rich countries, is not, paradoxically, to raise local temperature but rather to keep it from falling. As heat leaks out of inadequate buildings we have to replace it by adding more. Historically we have always done so by using fire, in open fireplaces, in stoves, and - more recently, particularly in rich countries - in boilers for so-called 'central heating', circulating hot water through radiators throughout a house or other building. In an open fireplace, most of the heat from the fire, the burning fuel, usually wood or coal, escapes directly up the chimney. A stove, burning wood, coal or fuel oil, is somewhat more effective, depending on its design, but still loses a lot of heat up the chimney. A central-heating boiler, usually burning fuel oil or natural gas, may be no more effective than a stove. But modern designs, particularly those that burn natural gas, arrange the pipework so as to condense the water vapour in the flue gas, recovering the heat that vapourized it before it escapes out the chimney, a significant improvement in performance. Central heating also requires electricity, to run the pump that circulates the water. As always, the better the thing that does what you want to do, the less you need fire or electricity. In this case the better the stove or boiler, the less fuel you need to deliver the heat you want.

You can get yet better performance by using so-called 'cogeneration' or 'combined heat and power' - 'cogen' or 'CHP' - producing both electricity and heat from the same fire and fuel. A typical cogen unit has an internal combustion engine burning diesel or natural gas, turning an alternator generating electricity.

The hot exhaust gas of the engine then goes through a 'heat exchanger' or boiler which produces usable steam or hot water, either for, perhaps, laundry or showers, or in central heating. Instead of burning the fuel directly in a boiler, producing only heat, with cogeneration you get the electricity as well, effectively free, as a bonus.

Advanced designs of cogen now use not traditional internal combustion engines but microturbines or fuel cells to generate the electricity, using the hot exhaust for steam or hot water as usual. A microturbine is exactly what it sounds like - a small gas turbine in a housing perhaps the size of a domestic wardrobe, turning a similarly small alternator. A fuel cell is a first cousin of a chemical battery. It generates electricity by a chemical reaction that uses fuel but not fire. The fuel required is the gas hydrogen. The exhaust gas from the fuel cell is just water vapour, completely benign, not a pollutant at all. The question you then have to ask is where you get the hydrogen. The commonest way is by using a controlled fire. You burn coal, oil or natural gas without enough oxygen to consume it completely, and add water in the form of steam. The fire strips the oxygen off the water to leave hydrogen, mixed with carbon monoxide. You dispose of the carbon part and use the hydrogen as fuel. You can also make hydrogen by electrolysis, using electricity to split water molecules into hydrogen and oxygen gases. As usual, you then have to ask where you get the electricity. If the electricity is fire-free so is the fuel cell.

Rather than using fire for heating you can instead use electricity. The simplest but least effective way is just to pass the electricity through a thin coil of tough wire, so that the wire gets red-hot. In the UK a device of this kind is actually called an 'electric fire'. A slightly more elaborate device surrounds the hot wire with firebricks which soak up the heat and then release it gradually, a so-called 'electric storage heater'. These usually operate after midnight, storing heat when electricity is cheaper and releasing it

113

through the following day. Unfortunately they are warmest when you least want the heat, and vice versa.

A much more effective way to heat a building with electricity is to use the electricity to run a heat pump. Just as an air-conditioner pumps heat 'uphill', out of a warm indoors to an even warmer outdoors, you can use the process to pump heat from a cool outdoors into a warmer indoors. An 'air-source' heat pump, as its name suggests, gets its heat from the outdoor air. Air is not, however, dense enough to be a very good source of heat. A 'ground-source' heat pump has a network of pipe outdoors, buried perhaps a meter deep in the soil, collecting heat from the soil, which is dense enough to provide continuous heat without itself cooling down significantly. Ground-source heat pumps, despite the cost and hassle of burying pipe, are becoming increasingly common, particularly in northern Europe. If you have a convenient seashore, lake or river, water-source heat pumps also work, with no digging. If the heat pump uses fire-free infrastructure electricity, such as hydroelectricity, windpower or solar power, you get heating without fire. When active heating is essential for comfort, when outdoor temperatures are too low for even the best building to keep its interior tolerable, you may well resort to fire. But you don't have to.

What if you actually want to raise a local temperature, to make it too hot to touch? The first and still much the most important reason we want to do that is to cook our food. Cooking also dramatically widened the range of what we could eat, especially after we used fire to fabricate water-tight ceramic vessels in which we could boil roots and grains. In rich countries we still sometimes cook on an open fire - say a grill, spit or barbecue, or a gas ring. Indeed such cooking is one of the few ways we now commonly see a genuine fire. But we much more often use some form of physical thing, perhaps a cooker with an oven, using fire or electricity more or less invisible inside, or a microwave, once again using electricity. Most of the time what matters is what we cook in, the vessel containing the food. Some need more fire or electricity than others, depending on how you deliver the heat to the vessel. But your domestic cooking is not, in the main, a significantly wasteful way to use fire.

In rich countries the waste occurs much more in the increasing prevalence of processed food. Evidence suggests that we are less and less willing or able to prepare raw ingredients for the dinner table. Instead we buy food that has been prepared for us, either ready to eat as it comes from the packaging, or requiring only to be heated before eating. The industrial cooking that prepares our processed food is another major use of fire, at least as important as the cooking we do ourselves. The processed-food chain also uses fire in other forms of activity besides heating, including transport, chilling, and packaging.

As well as cooking, we also use active heating to make hot water, for washing, bathing and laundry. That applies to households, hotels, hospitals, schools and many other places, particularly those where humans congregate in numbers. The water once heated is often stored in large tanks or cylinders, ready for use when you open the hot tap. Because the water is much hotter than its surroundings, however, a lot of the stored heat will escape, unless tank walls and piping have thick insulation to control heat-flow. Better insulation of hot-water systems is one of the easiest, fastest and least expensive ways to reduce wasteful use of fire or electricity.

In poor countries, especially in rural areas, the women - always women - still cook on open fires. It is a major chore, and an unpleasant one. The smoke causes millions of deaths of women and children every year. This primitive arrangement cries out for better. Entrepreneurs now make and sell a variety of simple cookstoves that use firewood much more economically, and channel the smoke out a chimney. Some rural villages with domestic livestock, notably in China, have established communal biogas digesters, which convert animal dung and urine into a clean fuel gas for cooking and also for lighting, minimizing smoke and fumes. Other entrepreneurs, especially in low latitude areas with abundant sun, make and sell solar cookers that collect and concentrate sunlight to deliver cooking temperatures, avoiding fire entirely. Better things to cook with are gradually reducing the damage done by fire, and making preparing food less of an ordeal. But the change is much too slow.

We also use fire to raise local temperatures to manipulate materials. We smelt metals such as iron and steel from ores. We make them into desired shapes by casting them as hot molten liquids, or forging the solid metal when it is hot enough to be soft and malleable. We weld metals with searing flame from welding torches. We roast or 'calcine' limestone and other materials to make cement. We bake bricks in kilns. We boil petroleum to distill or 'refine' it into separate fractions for different uses. We then combine some fractions as hot liquids with other materials to make plastics and other petrochemicals. For all these activities we get the necessary high temperatures from fire. To feed the fires of industry we have long used coal, for instance to make iron and steel, cement and bricks. For petroleum products and petrochemicals the fires burn some of the petroleum in order to convert the rest. Welding torches may use liquid fuel such as acetylene, burning in pure oxygen to make the flame yet hotter.

Some welding, however, uses not a flame but an electric arc. Another exception to the dominance of fire for manipulating materials in industry is the metal aluminium. We get it from the ore bauxite by electrolysis in vast smelters, any one of which may use the electrical output of an entire full-scale traditional power station. The power station may in turn use fire. But many aluminium smelters have been sited near major hydroelectric stations, often thousands of kilometers from the site of the bauxite mine. Cheap electricity compensates for the cost of shipping the bauxite. The smelting pots cannot be allowed to cool down. That would destroy them, as their contents solidified. Because of its origins, aluminium is sometimes called 'solid electricity'. Although it is also one of the easiest metals to recycle, we use it in vast quantities for short-lived purposes such as beer cans and processed-food containers that end up in landfill.

All these activities operate at temperatures so high that heat can readily escape to the surroundings. Insulation of hot industrial plant is even more essential than insulation of the buildings humans occupy, not only to keep down the cost of fuel to feed the fire but also to keep the surroundings habitable for workers. Yet industries often skimp insulation for reasons akin to those that have left us so many inadequate buildings. When the plant was built fuel was cheap. Insulation is an additional cost, high-temperature insulation especially so. It may not fit conveniently into the space available. Shutting down the plant to install it would be too expensive. Despite abundant evidence that adding industrial insulation will rapidly pay for itself, far too many industries do not bother.

Much of the active heat used in industry, moreover, is not for the industrial processes involved but simply to keep the industrial buildings tolerably comfortable for the humans inside. Industrial buildings such as warehouses are notorious for insulation that is poor to non-existent, requiring continual replacement of the heat that leaks out. Yet again, controlling heat-flow with better materials and structures would help to reduce the extravagant use of fire to compensate.

Some industries have recognized that they rely too much on fire, and use too much of it. Research has been under way for years to develop, for instance, alternative processes to manufacture cement. As yet, however, traditional fire-based processes prevail. But the potential is clearly there. The American visionary Amory Lovins likes to point out that we know three ways to make building material out of limestone. You can cut it into blocks. You can calcine it at high temperature, to make cement. Or you can feed it to a chicken. Weight for weight, eggshell is a very strong

117

material. But we do not know how the chicken does it. Moreover, it does it at its body temperature, barely warmer than the chicken coop. High temperature is not necessary. In the science of manipulating materials we still have a lot to learn. Indeed, in everything we do, we need to learn how to do it as nature does it, without the extreme temperatures of fire.

Cooling better

Buildings that do not control heat-flow effectively leak heat in both directions, outwards and inwards. In bad buildings, just as we use active heating to keep indoor temperatures from falling uncomfortably, we now use active cooling - air conditioning - to keep indoor temperatures from rising uncomfortably. As world weather grows ever more extreme, we have to expect more heat waves, more severe and in places not used to heat waves. More and more of us, in more and more places, now use air conditioning to cope with them. Unfortunately most air conditioning now in use relies on fire, directly or indirectly, to operate. In effect we are trying to cope with heat waves by making them worse.

Since adding air conditioning costs money, both to buy and to run, a better strategy would be to add insulation first, as well as improved windows and doors, shading and exterior blinds, to block the inflow of heat from outdoors. However, even after upgrading the building, as outdoor temperatures rise you may still need indoor cooling. The commonest form of air conditioning is a so-called motor-compressor heat pump, essentially a common electric refrigerator turned inside out. Instead of pumping heat out of a cool box into a warmer room, it pumps heat out of a warm room into an even warmer outdoors. If the electricity to run the motor uses fire, your air conditioner is working against itself. If it is fire-free infrastructure electricity, the process makes more overall sense, especially city-wide and country-wide.

But an even better option is to use the other kind of heat pump, known as an absorption chiller. Rather than using a motor-compressor it uses a heat source to drive the cooling loop. The fluid in the loop, evaporating at one end and condensing at the other, absorbs heat at the evaporating end and gives it up at the condensing end. Depending on where you put it, you can use it for a refrigerator or for an air-conditioner. Its great advantage is that you can use any available heat source to drive the loop.

In industry, for example, any time you need cooling as well as heating, you can use an absorption chiller in conjunction with anything that produces a hot exhaust, perhaps a boiler or motor. If you have an electricity generator using fire, whose hot exhaust also produces useful steam or hot water, you call it cogeneration. The same arrangement, producing both electricity and usable heat, but also using the remaining exhaust heat in an absorption chiller, you call trigeneration. Trigeneration can almost triple the amount of use you get out of the fire.

Sunlight too produces heat. If you use sunlight as the heat source for an absorption chiller, you can perform a kind of engineering judo - the hotter the sun, the better the cooling. The opportunity this offers, especially in sunny low-latitude countries growing ever warmer, may prove invaluable.

You can also deliver cooling as you deliver heating, through a network of pipes, either within a single building, central cooling like central heating, with chilled water rather than hot water, or over a wider area - district cooling like district heating. With suitable design you can use the same network of pipes to deliver either heating or cooling as needed, according to the season. As always, however, the first priority is to get the individual buildings right. In many places that will be a better investment than an extensive and expensive network of pipes, and make the pipes both uneconomic and superfluous.

A more vigorous form of cooling, dropping the temperature well below that of the immediate surroundings, we call refrigeration. We use it particularly to preserve perishable food, in the kitchen, in the food shop and throughout the chain of processing and delivery from producer to consumer. We mostly do it using the same physical processes in the same kind of devices we use for air-conditioning - heat pumps, either motor-compressor or absorption chiller. The device pumps heat out of the enclosed space we want chilled. Whether it is a domestic refrigerator or deep-freeze, a butcher's cold room, a cold-storage warehouse or a refrigerated truck, railcar or ocean freighter, the most important part of the system yet again is not the heat pump but the insulation. If the insulation is good enough, once the space is chilled the heat pump needs to run only intermittently, to remove any heat the insulation still lets in.

The fastest-growing category of cooling, however, has arisen only within the past two decades. The server-farms, the vast banks of computers that store and transmit data for the internet, use impressive quantities of electricity, and turn that electricity mostly into heat. Unless you remove this heat as fast as it arises you face the possibility of frying the microchips that manage the data. Chip designers strive to make them run as cool as possible, both to save electricity and to minimize heat output. But cooling is still a critical requirement for a server farm. Some recent farms have even been sited in northern latitudes and in caves, to make the outside temperature as low as achievable, and reduce the demand on the cooling system. Many of the earliest server farms relied on conventional electricity, much of it based on fire. But some of the most recent have opted for solar electricity to run their cooling, and absorption chillers with solar heat as the operating devices, avoiding fire completely.

As the planet continues to warm, cooling is going to become ever more essential, for food, for electronics and for people. But using fire to run a cooling system makes little sense. From now on, for cooling, we have to emphasize fire-free infrastructure electricity, absorption chillers with solar heat, and - above all - the best available insulation.

Lighting better

The best light is daylight. We humans evolved with daylight. Some scientists suggest that acquiring eyes, organs to respond to daylight, was what triggered the rapid burst of evolution we see in the fossil record, that eventually led to us and our many myriad fellow creatures. Yet those of us fortunate humans in rich countries now commonly spend less time in daylight than we spend in the light that we ourselves have made. Lighting better means not only making light better but using light better. That has to start with using daylight better.

What do we mean by 'using daylight'? Humans are diurnal. Most humans sleep during the hours of darkness and wake in daylight. We think of that as normal. We even change our clocks to give us 'daylight saving' at more convenient times, as the seasons change and daylight increases and decreases. But countless millions of us then spend most of the daylight hours indoors, in buildings whose design all too frequently makes little or no use of the daylight surrounding them.

The problem starts with the trade-off between controlling heat flow and controlling light-flow. Tradtionally, to make a building heat-tight, you also made it pretty light-tight, minimizing light-flow as well as heat flow. Traditional glass windows, for example, even if well-fitting, which most were not, conducted heat all too well. To limit heat-flow you therefore wanted as little window-space as possible. Minimal window-space meant, in turn, little daylight indoors even in daylight hours. The advent of affordable electric light, however, made that much more acceptable.

Then came double-glazing, with insulating air-space between inner and outer panes, a definite improvement although often oversold. Its significant advantage for controlling heat-flow was that you could fit it on an existing building with comparative ease. But you could not so easily enlarge existing windows or create new ones. Double-glazing did not help daylighting indoors.

Throughout the twentieth century, in rich countries, electric light made daylight more and more disposable. Architects assumed electric light twenty-four hours a day throughout the year, indoors and even out. Daylight even became a major problem. The rise of so-called 'curtain-wall' office buildings, completely sheathed in glass, might prompt the unwary to assume that architects were at last using daylight. Unfortunately almost the exact converse was true. Curtain-wall buildings did indeed collect daylight. But its main effect was not to illuminate the interior, but to heat it, like a greenhouse. So-called 'solar gain' became a serious drawback of glass-walled buildings, placing severe demands on the air-conditioning that had also become an essential feature of these designs. By the late twentieth century the built infrastructure of cities all over the planet had become helplessly dependent on active processes using fire and electricity, not only for controlling heat-flow but also for lighting.

That will not be easy to rectify. But lighting offers ample potential for improvement, starting with daylighting. The first step is simply to recognize the importance and value of daylight inside a building. The next step is to invite it in, as welcome illumination rather than troublesome heat. You can get daylight into a building in several ways. The most obvious, if sometimes difficult, is to open the building structure to the outside. Refurbishing an older building, for example, now often includes creating an atrium - an open area in the interior, with wide window areas above and around it, admitting daylight that can then also enter the rooms surrounding the atrium. Two millennia ago our Roman precursors erected many buildings using this principle - hence the Roman name 'atrium'. Many recent buildings, notably hotels, incorporate a daylit atrium as an integral feature of the structure.

Another option becoming increasingly common, technically ingenious, versatile and not alarmingly expensive, is the 'light pipe'. It is an optical device that gathers daylight at its outer end, on the outside wall of a building, and channels it some distance into the building, possibly several meters, pouring the daylight into an interior space that would otherwise never receive daylight. You can sometimes install a light pipe in a building when any other option for daylighting indoors would entail major structural work.

We have always used windows to bring daylight indoors. The latest designs of window offer exciting additional options. Instead of the awkward trade-off that makes windows a problem for controlling heat-flow, you can now get innovative surface-coating for windows that admits visible daylight while blocking the infrared light that causes solar gain. An even more startling possibility is photovoltaic surface coating, a window that admits daylight while generating electricity. We are on the threshold of a whole new approach to using daylight.

Using the light we ourselves make also offers immense potential for improvement. Within buildings, architects and designers often place the sources of artificial light in fittings that waste much of the light produced, sunk into ceilings, behind baffles in walls or otherwise blocked from delivering the light where it may be useful. We also fail too often to recognize that we use light in different ways. Bright clear light in front of you for a task at hand is not the same as casual light in a corridor or gentle light at a restaurant table.

The colour of light, the 'spectrum' of its component colours, is likewise important, not least psychologically. Because we are so used to daylight and its usual spectrum, we often find artificial light less comfortable, perhaps the blueish 'cold' light of some fluorescents, or the harsh yellow light of sodium-arc streetlamps. The early designs of light-emitting diodes, LEDs, suffered from this drawback, producing a spectral colour many found unpleasant. Recent designs of LED have largely solved this problem, producing much 'warmer', more congenial colours.

A more obvious problem is that we simply waste light. You may not think of it that way, but that is what we are doing when we leave on an artificial light we are not using. Daylight is free. Artificial light, light we ourselves make, is not. Yet anyone who ventures into an urban centre almost anywhere after sunset will be surrounded by buildings ablaze with light. Some is genuinely functional, lighting streets and pavements for pedestrians and vehicles, although all too often badly aimed, pouring useless light skyward. Some is thought to be functional, although that is debatable - company logos, shop windows, illuminated advertising signs and so on. But much city-centre light comes from the windows of unoccupied office buildings, sometimes more than fifty stories high, still lighted as artificially as they are in daytime, as if the daytime occupants of the building were still a ghostly presence out of office hours. The daytime occupants are not usually at fault for this waste of light. In many such buildings you cannot find a light switch, nor turn off an unused light. The building managers control the lights, and their default mode is 'on'. The occupants, needless to say, pay the bills.

Persuading office buildings to turn off unneeded lights at night would significantly reduce electricity use in almost every urban centre on the planet. That would, however, be an unwelcome development for many electricity companies. Wasted light at night helps to keep power stations operating when they might otherwise have to be shut down. It also provides the companies with a valuable source of revenue, since someone has to pay for the wasted light.

A similar corporate problem has long complicated the business of making light. From 1924 to 1939 major companies producing traditional incandescent lamps, in designs dating back to Edison, collaborated in a cartel, ironically known as Phoebus after the Greek god of light, not only to fix prices but to ensure that lamps burned out quickly enough to require frequent replacement. These companies had major investments in factories and assembly lines turning out such lamps in their billions, well into the 1990s. They were not keen to see these investments written off, in their eyes, prematurely. But Chinese entrepreneurs began producing low-priced compact fluorescent lamps, CFLs, that used far less electricity, delivered comfortable light and lasted far longer than incandescents which burned out after a few months. The Chinese began to export these CFLs to the western markets hitherto dominated by traditional incandescents. In response the western manufacturers tried to have their governments ban the import of Chinese CFLs, or impose high tariffs to make them much more expensive. In due course the opposition failed, and the western companies began frantic retooling to catch up with the Chinese.

The episode illustrated vividly a key issue that arises everywhere in human activities. Every time you find a way to make an activity better, someone will be a loser; and the loser will try to stop you. As LEDs gradually make inroads into the market for lamps of every kind, the manufacturers of CFLs will doubtless be doing their utmost to impede the change-over. They will not succeed; but they will assuredly slow it down.

Nevertheless we have huge potential for lighting better - using daylight better, and using the light we make better and wasting it less. We don't even have to do it ourselves. Sensors and controls can now detect when we need light and when we don't, and switch lights on and off accordingly. You have probably seen one of the famous photographs taken by satellites, of the areas of the Earth illuminated at night, the clusters of light in Europe and North America, the thinner bands in South America, Australia and Asia, and the dark patches, especially in Africa. On one hand, the images give a vivid picture of those who have and do not have electricity. On the other hand, they also demonstrate that those who do have are extravagantly wasting it. Why are we sending light out to be photographed by satellite? Why are we lighting outer space? This waste light, this light pollution, fills our night skies. As a result the children in rich countries now never see one of the most awe-inspiring sights of our human experience, the Milky Way, our home galaxy. We owe it to them to light better.

Exerting force better

Somehow, a long time ago, some Neanderthals decided to bang two rocks together. At least one flaked off enough chips to leave an edge you could cut with - a stone axe, perhaps the first tool. The force the Neanderthal exerted, no one now knows why, was an early example of another key human activity. We exert forces on ourselves all the time, our muscles moving our limbs and so on. But exerting force on other things can have several significant outcomes. Like the Neanderthal, we can use force to alter the shape of things. With the edged stone tool the Neanderthal could use force to hack the skin off a mammoth, and hew pieces of meat small enough to bite, chew and eat. We now use force with knives, forks and other utensils to prepare and eat our food. We use force with uncounted billions of tools, some simple, many unimaginably complex, to shape the artefacts with which we have long surrounded ourselves. We also exert force to make some of these artefacts move, tools moving other tools to create yet more artefacts - think of drills, saws, lathes, turbine blades and so on. However, we also use force to move ourselves, our goods and our artefacts, more and more incessantly, over ever longer distances. That use of force, to create mobility, is now so important it is a whole separate category of human activity, to consider separately.

Nevertheless, exerting force to shape things and make things is also now a vast and varied human activity, at every scale from fabricating microchips to removing mountaintops. Human muscles still play a key role, for instance in assembling delicate electronics, or stitching fabrics for clothing, work now frequently done in factories in Asia, by skilled workers who earn meagre wages in often trying and even dangerous conditions. But we now mostly augment or replace muscles by machines that use either fire or electricity.

We still use fire to exert brute force. The most violent way is with an explosive such as dynamite or TNT, that burns so fast its hot expanding gases create a severe shockwave, useful for instance in mining and also of course in warfare. For brute force under more control we also use fire, although much less often now with steam engines, more probably with internal combustion engines burning diesel, for instance, in earth-moving equipment. To exert more precise, focused and controllable force, however, we now usually prefer electricity in electric motors. In many manufacturing activities, for instance, we now use robots to exert force, driven by electric motors and guided by electronic management of the requisite information. The human involvement comes beforehand, in designing and programming the robots. Even fabricating them entails, perhaps, other robots, likewise designed and programmed, back up the chain until you get to the humans that fabricated the first robots.

How might we exert force better? Start by asking why we're exerting force. Does it need to be brute force? Could it be more controlled, better focused, more nicely judged, to do exactly what we want to do and no more? Many people might now argue that most of the undertakings that use lots of brute force, such as large dams and river diversions, mountaintop-removal coal mining, tar sands extraction or uranium mining, are at best ill-advised and frequently seriously destructive. Closer examination suggests that such undertakings proceed only because those promoting them do not pay the costs of the consequences of their actions.

But less extreme uses of force may also benefit from closer examination and critical analysis. In industry, for instance, we use enormous numbers of motors to drive fans and pumps that propel fluids through pipework. The amount of force required depends not only on the fluid, such as how viscous it is, but also on the layout and configuration of the pipework. If the pipes are narrow, with many bends, the force you need to propel the fluid is much greater, because the fan or pump has to work against the friction from the pipe walls. Making the physical hardware, the pipework, better, with wider piping and fewer sharp bends, reduces the amount of force required. By extension, that also reduces the size of fan or pump you need, and the amount of electricity the fan or pump requires to run. If the electricity is produced with fire, improving the pipework reduces the amount of fire we use.

With electric motors, computers, sensors and controls we now can have ever more sensitive control of the forces we exert - how strong the force, where, for how long and so on. We no longer need, for instance, to install oversize motors 'just in case', wasting investment and running cost and - as a corollary - wasting force. By getting the entire system right, we can minimize the waste of force just as we can minimize the waste of light. Yet again we find that the key to getting fire back under control lies in upgrading physical assets and infrastructure - better things. That will entail investment, but the payoff will usually be swift and continuing. So will the benefits.

Moving better

We *Homo sapiens* have been on the move for at least 70 000 years, ever since we set out from central Africa to spread across the planet. As a species and as individuals we are now moving more than ever. Sometimes we want to move, notably for recreation and leisure. Much of the time, however, we move because we have to. Within the past century, particularly in rich countries, we have laid out our activities ever farther apart from each other. We reside in one place, work in another, shop in yet another, play somewhere else and so on - and these different places are now so far apart that we can no longer walk or even cycle from one to another. The forces of our own muscles no longer suffice. We need the help of engines and motors, using fire or electricity.

Moreover we now make and produce what we need or want, our food and drink, our raw materials, our manufactured goods, even farther away, often on an entirely different continent. Getting them from where we produce them to where we use them now requires a relentless flow of movement by road, rail, sea and air. The fires in the engines providing the force to propel all this movement need continuous feeding. Moving everything, ourselves and our goods, now entails what is effectively a continuous raging fire across much of the planet. The consequences, for the air in cities, for security and international tension, and for what is happening to world weather, grow steadily grimmer. If we are to get fire back under control, we have to devise and implement ways to move better.

We can start by asking why we move, what we move and how we move now. Then we can consider options. Can we avoid moving at all? Can we move shorter distances? Can we move smaller quantities? When we have to move, can we make moving cleaner, safer and less damaging, locally and globally? Can we make moving rely less on fire? The answer to each of these option questions, depending on the circumstances, is Yes, we can.

Moving people we can call travel. Moving goods we can call transport. In either case the better option is to move less - less often and less far. That may mean rearranging our activities so that we don't have to move so much. Travel can be for business or for leisure. Business travel may be essential to establish personal contact with colleagues. But today's media for communication - audio and video phone, email, video conferencing, social media - make much subsequent travel unnecessary, incidentally also saving a great deal of time and expense.

When 'work' involved serious physical effort, with human muscles, you had to be at the workplace, perhaps to combine your muscles with those of fellow workers, to plant and harvest crops, to fell and chop firewood, to dig coal and ore, to construct cars and appliances and so on. In rich countries nowadays, however, more and more physical work of this kind is done by robots and other manufacturing technology. In rich countries, more and more of the work that people do is manipulating information - what we used to call paper work, more likely now to involve computers and networks. To do this work together we don't need to be in the same room, in the same building or even on the same continent. But we have developed ways of working, and the accompanying assumptions, that take for granted the need for co-workers to be in the same place - the office. You get up each morning and travel to the office to join your colleagues. It is of course a social occasion, and can be a welcome one. But for what office work now usually entails, the social gathering need not happen every day, or between the same times of day. The future of what we call 'work' is now a major and controversial issue in its own right. How it evolves will also affect profoundly how much and how far we'll have to travel. Business travel could become much less significant.

Leisure travel, by contrast, may continue to increase, for social and other reasons. Some leisure travel, of course, is just visiting family and friends, or for local recreation. But much leisure travel, over longer distances, particularly in and from rich countries, has long been a form of escape, an interlude away from the daily grind, often to places with sunnier or otherwise more congenial surroundings, especially for workers doing heavy physical labour, often under relentlessly unpleasant conditions. As working conditions, at least in rich countries, have improved that is now less of a consideration; but leisure is still seen as distinct from work, with leisure travel as its corollary. Tourism is now a key source of revenue for many parts of the world, often those otherwise considered poorer in conventional economic terms. Tourism has its problems. But on balance tourism and the leisure travel it entails probably help to make the world a better place, making its many and varied people somewhat less alien one to another.

Transport, moving goods rather than people, is quite a different issue, and much more challenging. The goods we move include raw materials, food and drink and manufactured products. The raw materials we move include metal ores, crude fuels such as coal and petroleum, and minerals such as phosphates for agriculture and aggregates for buildings and other infrastructure. Raw materials we get where we find them, rarely where we want to use them. That means transporting them from their source to wherever we actually want them, often on a different continent. We do so with road, rail, sea and air transport, almost always using fire. The distance involved is determined by where the raw material originates and where we want to use it, a distance not easy to reduce. The amount of transport is therefore determined by how much raw material we want to use. That in turn depends on how well we use it, and especially on how much we waste.

Reducing the amount of transport thus depends significantly on issues unrelated to transport, such as how we process and use raw materials including fuels. The most obvious way to reduce transport of raw materials, and the fire it involves, is to reduce the amount of raw materials we use, especially the enormous amount we now waste. The same also applies to transport of food and drink and of manufactured products. For food, much of the waste arises in industrial preparation of food, especially in rich countries, where pre-processed ready-made food constitutes an ever-increasing proportion of our daily diet. In poor countries, however, food is more often wasted in inadequate storage facilities, which allow the contents to rot or be eaten by vermin. In either case one accompanying result is an equivalent parallel waste of transport and the fire that drives it, carrying food that never reaches any dinner-table.

A further problem is that existing arrangements for moving, for both travel and transport, no longer work well enough, even on their own terms. Countless people in cities around the world now spend hours every day almost immobile, bumper to bumper, in vehicles whose fires cannot move them because they are surrounded by other vehicles, equally immobile.

If we don't want to be forced to move so much, we should be trying to reverse some of the trends that now compel us to use fire in vehicle engines for even the simplest of daily activities. City and other local governments should be endeavouring, for instance, to recreate what we used to call neighbourhoods, in which you could leave your home and walk or cycle to the school, the grocery, the bakery, the bookstore, the pub and other amenities, without having to use a car. As a particular example, out-of-town shopping centres, inaccessible without some form of motor vehicle, should be discouraged. The beneficial social corollary would be that the many, often the majority, who now do not drive a motor vehicle - the poor, the elderly, the young - would no longer be excluded from everyday life. Unfortunately, however, over the last half-century we have poured so much concrete, and arranged our activities so completely around it, that changing back will take a lot of time and a lot of money. Moreover, powerful interests are quite content with the existing dysfunctional arrangements. Not only do they see no need to change, they are adamantly opposed to change. Fire has many committed defenders, not least those whose revenues come from feeding fire. We shall have more to say about that.

As we strive to minimize unnecessary movement, we can also reduce the use of fire when we do need to move. Different modes of motor-travel and motor-transport use fire very differently, depending on the type of vehicle, what it carries and what powers it. Moving better and reducing fire may mean switching from one mode to another, as for instance from a private car to public transit, such as bus, tram or subway - as long as the public transit provides a good service, moves you when you want to move where you want to move at a price you find acceptable, not by any means always the case.

For personal travel, two conflicting trends have been evident since the early 1970s and the first so-called 'oil shock'. The fuel efficiency of some motor vehicles has dramatically improved, enabling them to move much farther on a given quantity of fuel and the fire it feeds. At the same time, however, initially in the US and then much more widely, a new category of motor vehicles has emerged, what in the US are called 'sport utility vehicles' or SUVs, larger than traditional cars, heavier and with much poorer fuel performance. Differential taxation, particularly in the US, has also favoured light trucks. Both SUVs and light trucks are clearly useful over poor roads, and to carry substantial loads. But they now routinely appear on city streets in rich countries, carrying just a driver, often for journeys of less than a kilometer. Tax policy ought instead to discourage this seriously retrograde trend.

You can move better and reduce your use of fire by selecting a different design of vehicle. For a private car you can simply choose a conventional petrol or diesel model with better fuel economy, taking you farther for a given amount of fire. That also applies to larger road vehicles, including buses and coaches for people and lorries and trucks for goods. Companies with fleets of vehicles can choose petrol or diesel models with better fuel economy and make major savings on fuel bills. For a private car or a public bus you can modify the engine and fuel supply so that the vehicle will run on liquid petroleum gas, LPG, or compressed natural gas, CNG, both of which emit fewer unpleasant fumes when burning. Now you can also choose a hybrid that combines both an internal combustion engine and electric motors, substantially reducing the amount of fire you need to cover a given distance.

Some public transit companies are now experimenting with vehicles such as buses powered by fuel cells and electric motors. In a road vehicle you can make hydrogen in a 'reformer' next to the fuel cell, and feed it with petrol or CNG. If you want to avoid the fire of a reformer, you can carry a tank of compressed hydrogen. If the hydrogen is from fire-free electricity, you can drive without fire. Thus far, however, the cost of fuel cells remains discouragingly high. Despite many years of promise, accordingly, fuel cell vehicles remain a rarity, private cars especially so.

More and more car drivers are now choosing one or another model powered purely by electricity, from a large battery that makes up most of the weight and most of the cost of the car. Whether this reduces your use of fire depends as always on where you get the electricity - whether it in turn is produced with fire or not. Although the combination is still very much the exception rather than the rule, you can once again move yourself and your goods in a private vehicle using only electricity, like the first electric-car drivers more than a century ago. As costs come down and battery-charging facilities become more widespread and more efficient, electric cars may become not the exception but the norm. They may then come to play an important role in electricity systems, as their batteries become a valuable form of dynamic storage on the system. We shall return to that shortly, when we describe and discuss the dramatic changes now altering the electricity systems around us.

A yet more striking development is also now under way. Some public transit systems already operate fully-automated driverless trains, controlled by computers and unmanned, and we use them without a qualm. More unexpected, however, is the prospect of driverless road vehicles. Recent research suggests nevertheless that this is not just a possibility but an inevitable eventuality. The dismaying toll of death and injury from road accidents is almost always the result of human error. Sensing and control techologies are evolving so rapidly that recent work on driverless vehicles has progressed much faster than its initial proponents anticipated. Electric vehicles will lend themselves much better to these designs and systems than will vehicles using fire, yet another stage in the shift from fire to electricity.

Cars without drivers may be coming. Cars without owner-drivers are already here, and becoming steadily more common. One of the disconcerting attributes of the private car is that in most instances it is in active use for at most two or three hours out of twenty-four, and often much less than that. If you own a car you not only drive it, but have to pay for fuel, and have to find and frequently pay for a place to park it when you're not using it. You also have to pay for maintenance, licence and insurance, even when you're not using it. In reaction to this, a number of cities now have car-sharing schemes, in which you pay only while you are using the car, taking it from your starting point and leaving it at your destination, to be taken in turn from this second location by another part-time driver, and so on. Each vehicle is thus in use for much more of the time, and you no longer have to worry about parking it. Combining the ownerless car with the driverless car looks likely to be an obvious corollary, especially with electric vehicles, with charging stations at the pick-up and drop-off points.

Nevertheless, even with clean vehicles and minimal use of fire, the present layout of our human activities means that our travel and transport problems are going to get steadily worse. Only by reducing our movement and our need to move can we keep global society, especially urban society, from seizing up in terminal gridlock.

Managing information better

We used to manage information just by talking and listening, with our own muscles, organs and brains. Then came writing and reading, in which we also used physical things such as rocks and chisels, styli and clay, parchment, paper, brushes, pens and ink, and eventually machines, including the printing press and the typewriter. We also developed calculation and measurement, with numbers of various kinds, counting procedures, arithmetic, geometry and devices such as the abacus and the calculating machine. In the 1840s Charles Babbage and Ada Lovelace dramatically expanded the concept of a calculating machine with Babbage's 'Difference Engine', yet another mechanical device, the precursor of what would become, a century later, the computer.

After dark we used fire to see what we were doing, but apart from smoke signals by day and beacons by night fire had no connection with managing information. Then came electricity. With electricity through wires, cables and in due course 'wireless' - radio - we learned to communicate and exchange information over astonishing distances and almost instantaneously. We also learned to record, store and recover information not only as writing, but as vision with photography, then as sound with the phonograph, then as both, with moving pictures soon including sound.

Early cameras and sound recorders were hand-operated or clockwork. They used neither fire nor electricity, except fire for flash photography and electric lamps and motors for movie projectors. From the 1920s onward, nevertheless, electricity became steadily more important as a means to record, transmit, receive and eventually analyze information for every human purpose. We used electricity as usual, to make light and exert force, for instance to operate movie cameras, projectors and record players. But a much more significant role for electricity was as the actual carrier of the information, along wires for telegraphy and telephone, by electromagnetic waves for radio and television, and eventually to store and process information - 'data' - in electronic form in computers.

The first modern computers used the electronics then available - vacuum tubes the size of incandescent lamps, and just as hot. Then came the transistor and its miniature manifestation, the microchip. In the 1960s a single computer filled a warehouse, its information input on punched cards. In the 1970s a mainframe was the size of a dinnertable. By the mid-1980s the personal computer sat on or under your desk, with a keyboard smaller than a typewriter. The internet and the World Wide Web linked computers, first in dozens, then hundreds, now in billions. Now, holding your smartphone you have vastly more stored information in your hand than the warehouse computer, and processing capability far beyond its imagining. With so-called 'quantum computing' on the horizon the trend is accelerating, and with it the use of electricity to manage the information.

Managing all this information better raises an array of issues. What do we mean, 'better'? Better how, and better for whom? Who has the information? Who gathers, holds and analyzes it? What do they do with it? We have entered the era of 'big data'. Retailers offer 'loyalty cards' that record your every purchase. Internet companies such as Google and Facebook track your activities and sell the results to advertisers. We have now discovered that our governments routinely intercept all our communications, our phone calls, texts and emails, and even track our whereabouts. They insist that they must do so to thwart criminals and terrorists. That puts you and me in the category of suspects. As the torrent of information engulfs us, the concept of privacy we used to take for granted is evaporating around us.

These pressing issues are now hotly debated almost worldwide, with no ready resolution foreseeable. But another aspect of the information revolution is easier to grasp. Storing, transmitting and analyzing all this information now represents perhaps the fastest-growing use of electricity. An ever-increasing number of data centres or 'server farms', some gigantic, each house up to hundreds of thousands of file-serving computers running continuously, all day and all year. They need not only the electricity to run the computers but also some way to cool them, to drain off the 'used electricity' turning to heat as it courses through the microchips. Cooling a server farm is now as important as providing the electricity in the first place. At least one imaginative entrepreneur is even using server-cooling to heat houses, turning what would be waste heat into a resource.

Although in our other human activities we have seen a gradual transition from fire to electricity, in managing information fire itself has never played a direct role. Behind the scenes, nevertheless, fire is still the way we generate most of the electricity we use to manage information. At the same time, however, the information revolution is helping to transform electricity systems themselves. That in turn will make managing electricity for information easier, cleaner and more effective. What we do with the information, however, will remain a challenge. Information you use becomes knowledge. Information you use wisely becomes wisdom. Will we use it wisely?

Better electricity

We fortunate ones in rich countries take electricity for granted. But more than a billion of our fellow humans still do not have electric light, or any of the other electric applications we use every day without even thinking about it. Moreover, those who make our electric applications possible are getting more and more uneasy. Traditional electricity arrangements that kept the lights on, at least in rich countries, for close to a century now look ever more precarious.

In any case we are at last realizing belatedly that traditional electricity leaves a great deal to be desired. It generates the electricity in very large, remotely-sited power stations, many of which however operate only intermittently or at only partial output most of the time, squandering costly assets. Large centralized power stations using fire waste two-thirds of fire's heat, discharging it unused. On many systems losses from wires waste much more. The arrangement, with its hundreds or thousands of kilometers of overhead wires operating in real time, is inherently vulnerable to disruption, by mishap or malice, over a wide area and almost instantaneously.

Traditional electricity assumes that every application is essentially the same, requiring the same high quality of electricity. This is all too much like the absurd water-management policy, in the UK and elsewhere. We collect rainwater, a resource that we receive everywhere, gather it up and centralize it in reservoirs, purify it to drinking-water quality, send it out through century-old pipes that leak, then use most of it for flushing toilets, washing cars and watering lawns. In the same way, we produce high-quality electricity as required by the most sensitive applications, such as microchip factories or paper mills. Then we use much of this high-quality electricity for undemanding activities such as heating and cooling. Most electrical applications are inherently intermittent or variable, as we switch things on and off. But large generators using steam from fire or fission are inherently inflexible. The arrangements for traditional electricity that we long took for granted are almost a total mismatch - even before we begin to notice the damage we are doing using fire in power stations.

Better electricity therefore should be cleaner and safer; more reliable and resilient; cheaper, by accurate accounting of all costs, including environmental; flexible, versatile and easy to control, preferably locally; and available to all, everywhere - a tall order. Fortunately the necessary transition is already under way. Unfortunately, however, it is happening too slowly, and in too few places. We urgently need to speed it up and spread it around, eventually worldwide.

The transition to better electricity entails better technology, better business arrangements, better finances, better regulation and other institutions, better psychology and better sociology - all in all a total transformation of what we expect of electricity and how we get it. That of course is why it is as yet happening only gradually, and only here and there. It is a major upheaval whatever your criterion, and the outcome is by no means assured. But the transition is steadily gaining momentum and gaining adherents, as its potential becomes clearer and its benefits more conspicuous.

Consider first the hardware to generate, deliver and use electricity - the system of physical artefacts with which we do what we do. We now have an extraordinary catalogue of available components to choose from, and it continues to grow. For applications and uses the list is especially impressive, particularly for managing information. Information flow and the tools to foster it, such as smartphones and tablets, social networks, blogging and vlogging, have expanded dramatically, and the expansion shows no sign of slowing down. For more long-standing applications, such as heating and cooling, illumination and motive power, the trend has been to better performance, partly by straightforward innovation and partly in response to regulation, as seen for instance in the ongoing transition from incandescent lamps to compact fluorescents, CFLs, to light-emitting diodes, LEDs.

Transitions always upset those left behind. Some users even insisted on being allowed to continue to use incandescent lamps, the century-old design turning almost all their electricity into heat rather than the light they presumably wanted. Manufacturers of refrigerators and deepfreezes also made a loud fuss about European plans to require performance labels on their products. But the outcome was that performance improved so dramatically that the top-line 'A' rating had to be amended to add 'AA' and 'AAA'. Washing machines, tumbledryers, vacuum cleaners and other household electrical appliances have shown similar improvement, as have household electronics including televisions, video recorders and computers. On the other hand some households in rich countries now, for example, routinely have multiple televisions and game consoles, including display screens almost the size of billboards, using more electricity than ever. Consumer electronics also have standby modes - critics call them 'vampire modes' - that drain electricity in astonishing aggregate quantities if you don't turn them off at the wall.

To generate this electricity we now have a panorama of different options, ranging in scale from a coal-fired or nuclear power station that can light an entire city, down to a pocket-sized solar charger for your smartphone. In many places, however, the huge power stations once taken for granted as the optimum are now an endangered species, like dinosaurs being overtaken by small agile mammals. In countries with elected governments, a nuclear station is now such a risky investment that no private finance will touch it without an ironclad guarantee of lifelong support from the government - you and me, as taxpayers. Coal-fired plants, too, are finding life suddenly difficult, as politicians at last notice the costs of fire. Some nuclear and coal-fired plants are still under construction, notably in China, where government - taxpayer - support is the norm. But even in China giant traditional stations are viewed with increasing discontent, as better options surge forward.

Some still favour fire, in the form of combined-cycle gas-turbine stations, CCGTs, which swept the board in the 1990s as the sudden preference for central-station generation. A CCGT station is cleaner, easier to site and much swifter to build and commission than traditional coal-fired or nuclear plant. In recent years, nevertheless, even CCGT stations are proving less than ideal, because of the new emphasis on truly clean generation, avoiding fire entirely.

For smaller-scale generation the commonest choice, once again, is fire, in the form of diesel-powered generators. We use many millions of them as back-up and standby units in case the traditional supply fails, as it does increasingly often even in rich countries. We also use them on building sites and in other workplaces, as portable machines easy to move from job to job. But diesel generators are noisy and smelly, and require expensive fuel, a particular problem when they are used, as they often are, in remote rural areas with no other electricity supply. Moreover we now know that the exhaust emissions from burning diesel fuel, especially the fine particulates, are even more pernicious than those from petrol.

When we use fire for electricity we can make better use of it by also capturing the heat. We can use the exhaust from internal-combustion generators such as gas turbines or stationary gas engines burning natural gas to deliver combined heat and power or cogeneration. Using the heat that would otherwise be wasted makes obvious sense wherever you also need heat itself, as either steam or hot water. Conversely, if you need heat and want to get it from fire, you can get more value from the fire if you use it first to generate electricity, and then use the heat as you desire. Many buildings, particularly in Europe, now have gas engines in the basement, generating electricity as well as hot water and central heating. Such a system requires no onsite maintenance. Instead the system is often instrumented with diagnostic software connected to a nerve centre supervising many such systems remotely, possibly even from another city. If the software indicates any abnormality, the nerve centre can dispatch a technician to rectify the fault, usually before the building occupants even notice anything wrong. Extending such an on-site system to include also an absorption chiller, to produce cooling for refrigeration and air-conditioning, gets still more use from the same fire.

Nevertheless, it is still fire, and still part of our problem, local and global. Better electricity must therefore aim eventually toward electricity no longer dependent on fire at all. Fortunately the options for fire-free electricity are burgeoning. They grow more numerous, more varied and more attractive by the day, offering a portfolio of different attributes that complement each other. They also come in widely differing sizes. You can install them in a large array, for instance as a windfarm on or offshore, grouping many units together, to produce collectively the output of a conventional fire-based generator. Or you can install a lot of single units in convenient locations, for instance PV on rooftops.

Either way the construction time is likely to be much shorter than that for a conventional power station based on fire or fission.

Fire-free electricity of course has been available ever since Volta's first battery, for which fire played no part in making the electricity, although it was essential for making the glass, copper and zinc. In the early years of electric light, water turned more dynamos than did fire and steam, and hydroelectricity still remains the largest source of fire-free electricity. But large hydroelectric dams share many of the drawbacks of large power stations using fire or fission, including unpredictably long construction times, cost overruns and major business risk. They also create massive environmental impact, flooding and obliterating landscapes, and may even trigger earthquakes. More attractive options now include much smaller hydroelectric units, particular those that need no dam at all, relying on steep terrain such as that in Nepal, where thousands of small hydro generators abound. Other small units, known as 'run-of-river' generators, simply sit in the stream, turned by the momentum of the moving water. Mini-hydro and micro-hydro generators can capture the natural force of moving water in many places where large hydro units would be impossible to site or finance.

Moving water is yet more tempting offshore, where wind makes waves of enormous force. But its very magnitude makes capturing the force of waves an engineering challenge that has thus far proved too daunting for commercial success, despite many varied efforts. Tides are both more predictable and more manageable than waves, leading to several successful generators of tidal power in suitable locations, notably in France and Canada. But suitable locations, with enough tidal range to be useful, that can be channeled through water-turbine alternators, are too few for tidal power ever to be more than incidentally valuable.

Wind, too, was turning dynamos in the early days of electricity, and yet more so in local systems in the 1920s and 1930s. The expansion of central-station generation, heavily subsidized by urban electricity users through the 1950s, wiped out many small rural local wind-generators in Canada, the US, Scotland and elsewhere. Since the 1980s, however, wind-turbine generation has become a significant contributor to electricity supply in Europe, North America, India, China and elsewhere, and its contribution is growing rapidly. So are the turbines. Individual wind turbines producing enough electricity to power a small town are already in operation, with yet larger coming off drawing boards. Installations often group turbines into windfarms with outputs like those of a traditional central station, particularly when sited offshore, as is happening more and more. The latest development for offshore wind is floating turbines tethered to the seabed, potentially a much easier arrangement for siting, especially in deeper water. Offshore wind technology still faces challenges, from corrosion and from violent sea action, but it could become a key source of fire-free electricity.

Another source, at least in certain locations, is heat from inside the earth. You could not capture useful heat from a volcano, at least not without some uneasy moments. But the hot interior of the planet does emerge at or close to the surface in more manageable form in some places, notably for instance New Zealand, Italy and Indonesia. The heat in the form of steam has been used for many decades to generate fire-free electricity in some 'geothermal' power stations. But it may be more useful just as very hot water, for district heating, as for example in Paris. Efforts continue to extract geothermal heat from deep boreholes into hot dry rock, in Canada, the UK and elsewhere. Thus far, however, the payoff has been limited.

The most versatile form of fire-free electricity is that from sunlight, the source that makes the planet habitable. One way is just to use the heat of sunlight to boil water or some other fluid, to turn a turbine, as you do with the heat of fire. You do however need to collect a lot of sunlight to boil enough water to turn a turbine of a useful size. You can do so with mirrors, many hundreds of them, perhaps mounted with motors to follow the sun and reflect its rays onto a tall central tower with a boiler at the top. Several 'solar thermal' power stations based on this concept have been installed in Spain and the southwestern US, and others are planned. One advantage of the design is that you can store some form of hot fluid, for example molten salt, for many hours, so that the installation can continue to generate electricity long after sunset, even throughout the hours of darkness. But solar thermal stations require a lot of land for the field of mirrors or 'heliostats', and the heliostats have to be kept clean for maximum output. That conflicts with the optimum location for such an installation, in a sunny but usually very dry desert area, where water for cleaning is limited. Sand, especially if blown by wind, is also unpleasantly corrosive to both mirror surfaces and tracking machinery. The pluses and minuses of solar thermal electricity are still under debate.

You can also convert sunlight directly into electricity, a process called 'photovoltaics', or PV. The concept has been around in practical form for at least half a century. Early spacecraft drew electricity from solar PV panels deployed in orbit. At the time, the panels were so expensive that only a government agency such as the US NASA could afford to use them. Nevertheless they worked, proving the concept of direct solar electricity. Since the 1960s technical development of solar PV has been steady, and more recently dramatic, improving performance and bringing down costs, until today, at least in some sunny locations, solar PV is already cheaper than traditional fire-based options, even without accounting for the true costs of fire.

Solar PV also comes in many varieties. You can install an array of solar PV panels covering many hectares of ground, that can produce as much electricity as a traditional power station. You can also take the same number of panels and divide them up, placing them in smaller groups on suitable rooftops, occupying space that would otherwise be unused, to produce as much electricity as the vast single array. You can install a modest array of panels combined with a battery, using the panels to charge the battery, to give you electricity around the clock, independent of any wider network. You can install panels on fixed mountings, so that they always face the same way, the simplest and least expensive arrangement, but not the most effective. Alternatively you can install panels on motorized mounts that track the path of the sun, keeping the panels perpendicular to the rays to gather the maximum available sunlight - more complex and expensive but with better output.

Recent innovations are introducing further solar PV options. Some focus the sunlight falling on solar PV media, to increase output. Some capture both the sun's brightness, as electricity, and its warmth, as useful heat, in a form of solar cogeneration. Some offer transparent solar PV coatings for windows, admitting daylight while generating electricity. Some print solar PV-active surfaces on fabrics, offering the possibility, for instance, of solar PV tents for emergencies, that not only provide shelter but can also generate enough electricity for lights and mobile phones. Solar PV fabric would also be easier to deliver to remote areas, rolled up to carry. The speed of innovation for varieties of direct solar electricity is breathtaking, as one startling novel concept after another emerges. Some will undoubtedly fade from the picture. Others may soon be commercial and widely available. The successful forms of solar PV will alter profoundly the way we produce and use electricity, the way we pay for it and the way we think about it.

Better electricity systems

Most of our uses of electricity have always been local rather than centralized. Most lamps for light, motors for force and electronics for information, for instance, are usually small-scale and distributed throughout our houses, workplaces and other locations. Until recently, by contrast, we have produced electricity in enormous remote central stations, before sending it out and dividing it up among our local applications. However, innovative generating technology, from gas-engine cogeneration to rooftop solar PV, is also local and small-scale, and may indeed be quite close to the application using the electricity, even in the same building. The implications, for system operation, for business models, for revenue streams and for regulation, are already seriously disruptive to traditional assumptions and arrangements, and are rapidly growing more so.

Electricity system operation has never been simple, because electricity is a continuous process happening simultaneously in real time over what now may be an entire country, and even across country boundaries. Even so, innovative small-scale generation adds a whole new degree of complexity. It may feed electricity into the network in places that used only to send electricity out. It may make electricity flow in unexpected directions through circuits, including protective devices that do not expect electricity to flow 'backwards' through them. System operators used to having complete control over how much electricity to feed into the network are worried about what they call the 'unpredictability' of wind and solar generation.

Advocates of innovative electricity point out that the unpredictability of many small units is less of a problem to the system than the abrupt failure of a single huge generating set, an occurrence not uncommon in traditional electricity. Indeed traditional electricity actually keeps large generating sets turning in step with the system, synchronized but without generating electricity, so-called 'spinning reserve', precisely as emergency backup ready to take over immediately in case a huge working generator fails. Spinning reserve, usually burning coal, relies on fire, but gets no use from it except as insurance, extravagantly wasteful but an essential safety-net taken for granted in traditional electricity. In any case, system operators have always coped with the unpredictability of people turning applications on and off. As far as system balance is concerned, decentralized generation may be variable but it is no different from people and their variable applications. As local weather forecasting continues to improve, local fire-free generation may become more predictable than people.

What does this mean for business models, and for financing electricity? Traditional electricity almost throughout the past century was a regulated monopoly franchise. You and I were captive customers. If we wanted electricity we got it from our one local supplier. We bought it by the unit and paid what the supplier charged us, a tariff set by the government or a government regulator. In many parts of the world that arrangement still broadly applies, for good or ill, even when the one local supplier has serious difficulty keeping the lights on, as is often the case. In the 1990s, however, particularly in many rich parts of the world, governments threw out this model. They broke up the monopoly and invited generators to compete to sell us electricity as a commodity, in what they called an 'electricity market'.

The idea, however, had one crippling flaw. In a market, if you are the seller, your revenue stream still depends on selling your commodity. Until you get the price you desire, you can withhold your commodity. Electricity, however, is not a commodity. It is a process. You cannot store a process, nor withhold it from the market. If you are a generator, you can refuse to generate. But if you are selling electricity, not generating means not getting any revenue. When you do not generate, when your generator is idle, you lose the opportunity to earn money with it, and you cannot store that opportunity. It's gone. Since the introduction of the electricity market in the early 1990s, many market participants have lost their jobs, their companies and their shirts. More than two decades later governments are still struggling to make the market idea work for electricity, with success at best limited. In the UK, the process the government calls 'Electricity Market Reform' has been underway for more than four years. The reforms being implemented may effectively almost entirely eliminate the market.

The rapid emergence of small-scale local generation, especially rooftop solar PV, especially when combined with on-site battery storage, complicates the market issue yet further, perhaps beyond resolution. Small-scale local generation close to users is an ideal match for local networks and private wires. They can link generation and users in a local system or 'microgrid', potentially independent of the larger network and its remote generation, no longer dependent on any larger-scale electricity market. Information technology already available can make a suitable microgrid self-stabilizing, especially if it incorporates active participation not only from generation and batteries but also from applications such as refrigerators, freezers and air-conditioners, able to switch on and off as the system requires. You as user probably don't even notice now when your refrigerator or freezer is operating. Control technology can seize that opportunity, to run chillers when generation is abundant, and turn them off when it is not, while always maintaining appropriate temperatures in well-insulated appliances.

One development that would have gratified Thomas Edison is the re-emergence of direct-current DC electricity. AC triumphed over DC more than a century ago because AC allowed you to use a transformer to raise the voltage and lower the current in a wire, reducing the heating-losses and making possible long-distance transmission. Recently, however, a new category of devices called 'power electronics' has come on the scene. They function rather like the transistors in microchips, but instead of manipulating tiny electric currents they can switch the heavy currents coming direct from full-scale power stations.

You can use an array of power electronics to convert AC into DC and back again, allowing you to increase voltage and decrease current and vice versa, much like a transformer for DC electricity. With power electronics you can therefore produce high-voltage DC, HVDC, which is much easier to transmit than high-voltage AC, especially over very long distances. More and more electricity networks are now including HVDC links, particularly between different systems, such as the underwater HVDC cables now connecting the UK, France, the Netherlands and Scandinavia.

As well as HVDC for long distances, DC has also made a comeback at low voltages. In a typical office building, for instance, all the computers and other electronics, as well as fluorescent or LED lighting and even high-performance controls for electric motors, all operate on and require low-voltage DC. But mains electricity arrives from the network as AC at significantly higher voltage. At the moment, therefore, every computer, for instance, has to have a 'power pack' to convert the AC to lower-voltage DC. So does every other DC application. The power pack turns a lot of electricity into heat, wasting the electricity and needing cooling to keep the device from overheating. Entrepreneurs in the US have now begun rewiring office buildings to deliver low-voltage DC to all the on-site applications from a single on-site AC-DC converter. That also opens the possibility of using, for instance, the DC output of your own fuel cells and PV as a direct feed to your on-site applications, bypassing AC conversion completely.

An additional corollary now often mentioned is the potential role of electric vehicles, with their rechargeable batteries, as a form of storage on the electricity system. Batteries too require and deliver DC. If you have an electric car you will probably plug it in to charge overnight, helping the electricity system to make use of generation, perhaps for instance that from windpower, that would otherwise be wasted. Conversely, with appropriate financial arrangements, your car could deliver electricity into the system at times of peak requirements, helping to keep the system stable. That would of course also entail suitable AC-DC conversion electronics at the connecting point.

You may have heard a lot lately about 'smart grids' and 'smart meters'. Introducing power electronics and the latest sensing, control and information technology on large traditional systems may transform their configuration and operation for the better. On the other hand, it may reinforce the traditional centralized model, against the disruptive effect of decentralized local generation. The debate is now raging, about the meaning and function of 'smart' electricity. How smart is it, and for whom? What information does it gather, and what does it do with it? Is it for you and me as users, or for the operators and managers of the electricity system? Who benefits, and how? These questions are critical, and they are now being hotly debated.

Better electricity business

Perhaps the most striking corollary of recent developments is that you as a private citizen may now generate your own electricity as a matter of course, with no reference to central authority. Whereas before you might have had to rely on a noisy, smelly diesel generator burning expensive fuel, you now have the option of your own generation that is clean, quiet and fire-free, at an investment cost looking ever more attractive and a running cost effectively zero. That option is not yet readily available everywhere - but where it is, it is profoundly disruptive to traditional electricity. The existing business model for electricity, how we finance it and pay for it, is breaking down.

For most of the past century electricity was not really a business. With a monopoly franchise, and customers who had no choice but to pay what it charged, the electricity producer and supplier had a key role in society and the economy. Its activities, however, did not include 'doing business' as we commonly understand the expression. People called the local electricity supplier a 'utility', implying that it delivered a useful service to the community on a basis agreed with the relevant government. Indeed it was often a part of the government, local, regional or even national.

When, after 1990, free-market promoters abolished the monopoly franchise, introducing competition and what they called an electricity market, they thought they were turning electricity into a business. However, they expected the system to work in technical terms exactly as before. It still therefore needed the network, for which someone had to pay. In effect, despite the abolition of the franchise for generators, the network would continue to be a monopoly, and someone from the government would have to regulate that monopoly. In the new market framework, the regulator would impose charges when you used the network to carry electricity between buyers and sellers. In practice, despite the rhetoric of free-market enthusiasts, close to half the price of a unit of electricity was thus determined not by a market but by regulatory decree. More than two decades later it still is.

Some policy people nevertheless cite what they claim to be the cost of a unit of electricity from different types of generator, often

153

as precise as fractions of a penny per unit. They often assert, for instance, that large-scale remote fire-based generation is 'cheaper' than smaller-scale cogeneration or fire-free generation close to users. In this 'yay-boo' approach to planning, advocates insist that this purported cost advantage means you should invest in fire-based electricity for your new generating plant. This comparison, of course, ignores the pernicious and costly effects of fire. It also, however, ignores other critical factors. What happens if you compare and contrast the different accounting and financial frameworks for the different options? How does the tax treatment differ? What are the many and subtle subsidies, especially the extravagantly generous subsidies long given to fossil fuels and nuclear power? What are the financial and environmental risks, and who bears them? What are the system and network effects if you add this new generation? Without explaining and exploring these and other essential differences between generating options, such cost comparisons are meaningless. They should have no influence whatever on policy. Policy determines costs - not the other way round.

The same is true for prices. Patrick Moriarty, the admired head of Ireland's State Electricity Board, once summed it up concisely: 'The price of electricity is what the government wants it to be.' Recent controversy over the proposed Hinkley Point C nuclear plant in the UK underlines this blunt conclusion. As this is written, in order to persuade Electricité de France to build the plant, the UK government has agreed to guarantee to pay the plant operators twice the present wholesale price of electricity for 35 years. UK taxpayers and electricity users will foot the bill. Sometimes, it seems, cost comparisons, however convincing, do not after all carry the day.

A better business model will have to recognize that electricity is not a commodity but a process. We electricity users are not buying a commodity. We are buying access to the process, at a particular place and time. That has to be the basis of the

154

transaction. Moreover, from now on, as local generation burgeons, we are ever more likely to be involved not only as users of electricity but also as generators, contributing to the process as well as benefitting from it. How we identify and assign value to these contributions is going to be a key to a business model for electricity that satisfies all participants, as good business should. The revenue streams should follow the values, delivering revenue from those who benefit to those who contribute, wherever they are on the system. Better electricity must also be better business for everyone involved.

Fanning the flames

Better things and better electricity will be the two keys to getting fire back under control. That means that better electricity must eventually be fire-free electricity, and sooner rather than later. If you assess dispassionately the ever more pernicious effects of fire on the world around us, that statement ought to be uncontroversial. It is not. Some of the loudest and strongest voices in global society are vociferous and unrelenting in their defence of fire. That is not, of course, how they would describe what they think, or what they are doing. Like it or not, however, they are defending and indeed promoting fire and its ravages. They are fanning the flames, expecting and intending that fire will rage ever more fiercely across the world in coming decades. Those of us who want to get fire back under control had better realize that we are waging an unequal uphill battle against a phalanx of powerful foes.

If that sounds unduly dramatic, try enumerating those who want to keep on feeding fire, regardless of its consequences. Start with countries. Entire countries now depend on revenue from extracting, processing and selling fuel to feed fire. Think of Saudi Arabia and its fellow OPEC members for oil, Australia, Indonesia and South Africa for coal, Russia for natural gas - and those are only the largest exporters, with many more also involved. Think of companies - not only the international oil companies, such as ExxonMobil, Shell and BP and their competitors, but also the newly powerful national oil companies of the Middle East, China and elsewhere. As well as the exporters you also have the domestic suppliers, particularly domestic coal suppliers in dozens of countries, from China to the US to Poland and elsewhere, and the US companies newly flush with shale oil and gas, among many others.

Think of the politicians in thrall to the fire-feeders, their campaigning coffers filled with largesse in return for legislation, regulation and diplomacy that leaves fire unconstrained. Think of the manufacturers of cars and trucks, stubbornly opposed to any measure that might limit the noxious emissions from fire in vehicles. Think of journalists and broadcasters and their bosses, all too often dependent on advertising from the fire-feeders. Think of economists and other academics, whose models, analyses and commentaries take for granted what they consider to be the essential role of fire and its feeders in our human activities. Think of trades unions, so often determined to defend the dirty and dangerous jobs their members now have, instead of advocating cleaner, safer jobs no longer dependent on fire.

Think, especially, of global finance, and the vast sums of money it now deploys, on the assumption that fire will continue to rage worldwide. Your pension and mine may now be held hostage by investments that take for granted the value of coal, oil and natural gas still underground, but earmarked to be dug up, sold and fed to fire, to generate the revenue on which our pensions depend.

These various interest groups hold different views about fire, and support different kinds of fire, although they never actually mention fire itself and probably don't even think of it. But they all hold one view in common - a narrative about what we do and how we do it. They are telling themselves, and us, a story. They believe it, and they expect us to believe it too. It goes like this: in all our human activities we need something called 'energy'. We need energy to keep ourselves comfortable. We need energy to cook our food. We need energy to keep the lights on. We need energy for motive power and mobility. We need energy to manage information, to run our computers and smartphones and other electronics. The energy story does not, however, put it like that. It doesn't talk about what you and I actually do, or what we do it with. It talks about the domestic sector, the commercial sector, the industry sector and the transport sector, all rolled together into averages and aggregates of 'energy consumption'.

Governments and companies produce and deliver this energy, in the form of oil and oil products, coal, natural gas and electricity. We buy it, we pay for it by the unit, and we consume it. If we cannot obtain it we face an 'energy crisis'. If we cannot afford it we are in 'energy poverty'. Politicians and the media pontificate about 'energy security', by which they mean guaranteed affordable supplies of oil, coal, natural gas and electricity.

Governments, companies and academics gather and publish data on how much of this energy we consume, in different ways in different places at different times, and try to forecast how much we may consume in the future. Governments, companies, academics and international organizations draw up scenarios of how we shall be consuming energy twenty, thirty or fifty years hence. Governments, guided by their advisors, academics and energy company lobbyists, formulate and implement energy policy accordingly. Their aim is to ensure the requisite investment in production and delivery facilities, to provide citizens and customers with the energy they will want to buy and consume.

Even as they cling to this story, however, most governments and many companies do now acknowledge, if sometimes grudgingly, that consuming energy, as they put it, has unfortunate side-effects both locally and globally. They are, however, torn between the desire to mitigate these side-effects and the imperative, as they see it, to continue to produce and consume energy in ever-increasing amounts. As a result efforts at mitigation invariably fall short. Day by day the side-effects are growing ever harder to ignore, and the energy story is becoming less and less persuasive.

Some of those feeding fire are beginning to have doubts about this story. Other fire-feeders are redoubling their efforts to reinforce the energy story, using every method they can command. They insist that all our human activities are less expensive when we use fire, although that is not of course how they put it. They downplay or deny the effects and costs of fire, especially the global effects. They fund nominally independent scientists and institutes to cast doubt on the ever more alarming scientific evidence, that the fire we use is heating up the planet. They bankroll politicians, to thwart attempts to legislate, regulate or negotiate limits to our use of fire. With the help of journalists and broadcasters, television, radio, print and social media, they feed into the public discourse a steady stream of misinformation and downright lying.

One way and another, those defending and promoting fire have thus far succeeded in defining the language and concepts of this debate, not least by focusing on what they call 'energy' and simply ignoring 'fire', the word and the process. Instead, over the years, those concerned about the effects of fire have focused their attention not on fire itself but on its consequences, especially the airborne wastes it produces. Against the implacable opposition of the fire-feeders, governments have sometimes imposed limits and charges on emissions of smoke and ash, sulphur and nitrogen oxides, petrol and diesel fumes and carbon dioxide. Some advocate, thus far with little success, a tax on emissions of carbon

158

dioxide. However, by trying to deal with the symptoms rather than the cause, they are denying themselves the most direct measure to mitigate fire damage. If you want less fire damage, use less fire. That outcome, however, is what the fire-feeders fear - and what the rest of us, our cities and our planet, urgently need.

Firefighting

The energy story is plausible, so much so that it has prevailed for decades. But it omits the most important feature of our human activities, the physical things with which we do what we do. The better the things - buildings, lamps, heaters, chillers, motors, vehicles, electronics - the less fuel and electricity, the less 'energy', we need. Moreover the energy story fails to account for the ever more costly damage we are doing with fire. Knowing what we now know, we can no longer call fire the cheapest way to do what we do. Instead of fire we can use fire-free electricity, probably already cheaper than fire in real terms, and steadily getting cheaper still. But first we have to shake off the prevailing mindset. The old story, the energy story, is getting us into ever deeper trouble. We need a new story - a new rallying cry. Fortunately it is already taking shape.

What does it say? If you've read this far, you may already have a pretty good idea. In this new story no one wants 'energy'. We want comfort, cooked food, illumination, motive power, mobility, information and so on - what we can call applications or services. We get these services from our human-activity systems. These systems are made up of elaborate arrays of physical artefacts, things we have made - the aforementioned buildings, appliances,

159

lamps, motors, vehicles, electronics and so on. To deliver any particular service, the system, the array of things, may also use either fire or electricity or both.

We fortunate ones, particularly in rich countries, have long been able to rely on our systems to give us adequate services, at least most of the time. But they are becoming ever more vulnerable and fragile, and the side-effects ever more alarming. If we want to do better, our activity systems will have to deliver better services, and deliver them better. Our systems and the services they deliver should be clean, reliable, affordable and fair, and available everywhere, not just to us the fortunate.

How can we make our activity systems better? In this new story we start by focusing on the physical artefacts, the things we make to do what we want to do. The most important things of all are buildings, the structures within which we humans now spend most of our time worldwide. Report after report has shown that improving buildings can reduce rapidly and dramatically the amount of fire we need to use to be comfortable. That in turn reduces the running cost of the buildings, and makes our comfort less vulnerable to interruption of fuel supply. We already know how to make most buildings better, some of them much better. But we have to want to. Providing the necessary incentives and overcoming the obstacles, including initial cost, the hassle involved and the split interests between landlords and tenants, should be a top priority for policy and policymakers, much more important than any form of fuel or electricity supply.

Much the same applies to all the other innumerable, multifarious things we use in our activities. We can make most of them much better, and that should be our first objective in every case. The payoff from improving our things will usually be quicker, more substantial and longer-lasting than from anything we can do to improve our access to fire or electricity.

The second theme of this new story is to acknowledge and account for the true cost of fire in our activities. When we have a choice of options, and are comparing the costs of alternatives, we can no longer afford to give fire a free pass. We know and can see every day just how expensive fire really is, in human health, in unpleasant surroundings, in dirty and dangerous jobs, and in the havoc it is wreaking on the only planet we have. Let us stop pretending the fire option is the cheapest, when we have cleaner, safer options. Above all, let us stop pretending that we need to use fire to make our electricity. We do not. Switching from fire-based to fire-free electricity is already happening, in Germany, parts of Australia, parts of the US and in many other places, and the pace of change is accelerating, around the world. The faster we can make the switch, the better. Companies and governments that seize this opportunity will thrive. Those that try to thwart it may not survive.

As electricity changes, we are also bringing it closer to home. Traditional electricity, with its enormous remote central-generation power stations using fire or fission, and its thousands of kilometers of overhead lines, is steadily giving way to much more decentralized electricity, smaller-scale generation much closer to users, much of this generation fire-free. We are seeing the rise of microgrids, integrated local systems, locally owned, operated and optimized. The first priority is to get the things right, especially the buildings, so as to waste as little of your locally-generated electricity as possible. Microgrids include both generation and applications of broadly similar size, controlled and stabilized with the latest information technology. You can connect your microgrid to the larger system, for extra stability and backup. But if the larger system fails you can operate your microgrid without it, and keep your own lights on.

In this new story some activities are changing faster than others. For controlling heat flow, raising and lowering local temperature, making light, exerting force and managing information, we are already seeing a change to better things, and to fire-free electricity. For travel and transport, however, we cannot rapidly do without fire. The electric car is coming, but the electric aircraft is not. Priorities are important. We need to concentrate our efforts where the transition is easiest and the payoff quickest. That will win us time to grapple with the less tractable, with the activities we shall find hardest to change.

Unfortunately but inevitably, the other theme of this new story is conflict. Those of us who subscribe to this story recognize that as yet we are heavily outgunned by those defending the old energy story - defending and promoting fire. They still have the money and the power, even though the evidence mounting on all sides is against them. Those of us keen to foster change have to cope first of all with inertia. Legacy assets, the old things on which we have long relied, are still there, many still using fire and aggravating its consequences. Legacy institutions, including governments and regulators, companies, financiers and the media, still favour choices and decisions that support the old energy story. The legacy mindset still prevails.

That said, the opposition from those who fear to be losers is now meeting challenge from those who see themselves winners. The winners include, for instance, companies that specialize in improving buildings and their contents, a dramatically expanding business. Other companies manufacture, install, operate and maintain equipment for decentralized electricity, fire-free electricity and microgrids. One promising development recognizes that those parts of the world not already burdened with, for example, centralized electricity systems, such as rural areas in poor countries, can leapfrog ahead. Without having to overcome opposition from legacy interests, they can move directly to installing integrated optimized local systems, using fire-free generation for local applications for local people under local control.

Financiers and investors, watching the evolution of risks for different investments, are at last beginning to have doubts about fire and those whose revenue depends on feeding it. Even the Bank of England has begun to question long-standing assumptions about the eventual value of fuels to feed fire. Are they actually assets? Might they soon be liabilities? Insurance and reinsurance companies, forced to anticipate ever more extreme weather and the damage it may cause to properties in their portfolios, are having second thoughts about putting their clients' funds into investments that depend on fire.

The essential change that we need to foster is a change of mindset. We have to change the way we think about what we do. This clash of stories is a clash of futures.

Choosing the future

We have a choice. We can look at the massed ranks of those defending and promoting fire, shrug our shoulders at what they insist is inevitable, and give up. Or we can look around us at what is already happening, as more and more people challenge fire and the fire-feeders, and we can join in. We may not yet have the power, but our power is growing rapidly. Widespread polling tells us that we already have the numbers; most of us want change for the better. We also have the better story - more complete, more accurate and more exciting.

Policymakers trying to cope with the pollution strangling our cities, the reliability and security of our human-activity systems, and the relentless overheating of our planet are looking in the wrong place. Governments and regulators, diplomats and negotiators are preoccupied with symptoms, while missing the cause. Shifting focus to the cause - fire - might at last elicit progress.

When fire was easy, with fuel to feed it ready to hand and abundant, using fire made obvious sense. But fire is no longer easy. In poor countries fuel to feed fire is becoming ever scarcer or more expensive or both. You may have to walk much farther to gather it, or you may have to pay much more to buy it. In rich countries the fuel we use to feed fire comes from ever more extreme places, posing ever greater risks, financial, geopolitical and environmental. What now makes obvious sense is to reduce our dependence on fire.

To do so, we must meet fire head-on, at the point it claims to be strongest. We have been told for decades that using fire to do what we do, and especially to generate electricity, is the cheapest, the least expensive way to do it. That alleged advantage is still the key selling point for fire and its advocates. It is nonsense. You may not be able to put a precise number to it. But assessing the true cost of fire, to human health, to our surroundings in cities, to military budgets defending supply lines, and to the natural systems that keep our planet habitable, indicates that - far from being first preference, as it now is - fire should be a process of last resort.

That applies particularly to generating electricity. In the decentralized electricity now emerging, fire is already less important. If we still use fire, cogeneration and trigeneration in local systems doubles or triples the value we get from it. Fire-free electricity channels natural forces to raise and lower local temperatures, make light, exert force, move things and manage information, without extremes of temperature or pernicious waste. Generation and use are ever closer together, both in space and in size - local electricity for local people under local control. Local microgrids make wider interconnections less crucial and less vulnerable, reinforcing reliability and resilience. As electricity evolves from centralized to decentralized, it becomes more democratic, with local ownership taking local decisions. Decentralized electricity also means using local resources, harvesting natural forces locally, with no need to bring fuel from below the Arctic Ocean, Siberia, the Middle East or other demanding locations. Decentralized electricity epitomizes the growing power of local agendas for local people.

National governments continue to confront each other in bitter stalemate over measures to control fire. But local governments of cities and communities around the world are now cooperating to advance a common purpose - to clean up their air and water, minimize waste, reduce fuel and electricity bills, and improve the lives of their citizens. They may not describe their efforts as controlling fire, but that, too, is what they are doing, and we are all the better for it.

National governments, international organizations and non-governmental organizations have put forward a lengthening catalogue of scenarios and roadmaps to describe possible futures, for individual countries, for regions and for the whole world. They usually start by describing what they call 'business as usual'. They then alter various assumptions in their computer models, and watch the trajectory into the future gradually alter accordingly. The results can be instructive, informative and sometimes counterintuitive. However, what is happening now is much more radical than a gradual change of trajectory. We are not just changing the trajectory a bit. We are beginning a fundamental transformation of human-activity systems.

To speed the transformation, we should use more accurate language and concepts to describe what we do and how we do it. We should rescue the valuable word 'energy', and restore its original meaning, as scientists and engineers have always understood it. Energy is the unifying principle of the universe. Everyone knows the first law of thermodynamics, although you may not know it by that name. As the law of conservation of energy, it says that in any process whatever, of any kind anywhere, energy is never created nor destroyed. You do not need to conserve energy. The universe conserves energy. To degrade the profound physical concept 'energy' into simple shorthand for 'oil, coal, natural gas and electricity' deprives us of a key word for our new story and our better future. We should reclaim it.

We should look at the options, opportunities and potentials from a fresh perspective, including the urgent need to get fire back under control. We should focus our efforts where they will yield the best results - not on fire and fuel but on the physical artefacts and infrastructure, the things we use, and on fire-free electricity to drive them. In this context, the most important role for government is neither legislation nor regulation. The most important role for government at every level, civic and municipal, state and provincial, federal and national, is as a major participant in our human-activity systems. Governments are responsible for the buildings, fittings and appliances they themselves use; for street lighting; for data processing, communications and other electronics; for vehicles of every kind; for the police, the military, prisons - the catalogue is endless. Governments are major customers for the suppliers, not only of fuel and electricity but also of the infrastructure and other physical things governments use.

Governments make the rules. Governments, as major and desirable clients, can therefore redefine the business they do, upgrading to better things and decentralized electricity. They can call for tenders for detailed audits of their buildings and other physical properties. They can commission contractors to identify, design and implement improvements, such as better insulation, better doors and windows, higher-performance lighting and motors, and on-site generation of electricity, heat and cooling. Governments can integrate and optimize entire systems throughout their facilities. They can install, operate and maintain the upgraded systems, not merely ad hoc, as sometimes happens already, but as coherent strategic programmes with long-term focus. They can report on and publicize the progress and results of these programmes, to inform, educate and stimulate the private sector to seize similar opportunities. Such government programmes will prime the pumps for companies that invest in and operate whole systems, not merely supplying fuel and electricity. Because such government programmes must be implemented where the buildings are, they will create jobs all over the country, mostly where workers are keen to have them. Best of all, if managed effectively, such government programmes will save us taxpayers money.

Governments, however, have to want to take this lead in choosing a better future. Thus far, national governments seem to be laggards. But city governments are forging ahead. So are enlightened companies. Those who drag their feet will be left behind.

Beyond the Fire Age

The changes now transforming electricity, from centralized to decentralized, and from fire-based to fire-free, mirror changes that could dramatically alter both the global economy and global society. The implications reach far beyond fire. Apart from food, fuel to feed fire is the only product we make that we intend to be consumed continuously, to be used up, to be continuously replaced. Everything else we make - clothing, footwear, furnishings, tools, vehicles, buildings - is, or should be, durable, something that lasts. Since the 1980s we have even given this quality an ungainly name - 'sustainability'.

Despite such putative aspirations, however, we have instead created a global economy modeled on fire and its consequences, a 'consumer society' whose central function appears to be to turn resources into waste as fast as possible. As far as this global economy is concerned, you and I are consumers. Our role is to act like fire, to consume resources. The oxymoron 'consumer durables' succinctly pinpoints the paradox.

This is stupid and dangerous. We urgently need to move away from fire as the model for human activities. That in turn will entail changing the model of our global economy. We need to change the way the world works - a daunting but exhilarating challenge. To move beyond today's destructive Fire Economy we need different ground-rules. We need different forms of business and commerce. We need different transactions and business relationships, not short term but long term and durable. We also need appropriate regulatory frameworks, and financial instruments and practices. We need to rethink the whole value structure that governs what we do and how.

With the right ground-rules we could move to a global economy based not on short-term commodity transactions but on longer-term investments, and mutually beneficial relationships to match. Just as today's destructive short-term economy resembles fire, based on consumption of commodities, so the durable longer-term economy resembles fire-free electricity, based on process, access to process and services. We might therefore, perhaps, anticipate and work towards a transition from an economy model emulating fire to one emulating fire-free electricity. We could aim to replace the Fire Economy with an Electric Economy.

For the moment any such idea is speculative, to put it mildly. But it fits into our evolving understanding of what we do, how we do it and how we can do better. Our economic system reflects what people call our energy system - what I prefer to call our human-activity system, how we do what we do. If we are to change this system - and we must, to preserve habitable cities and a habitable planet - we have to change the economic and social system of which it is such a central part.

That is not as unrealistic as it may sound. Few would claim that the global system we now have is working adequately. Far too many of our fellow humans have little or no access to the services we fortunate ones take for granted. The yawning disparity between the wealthy few and the rest, even in rich countries, is not a recipe for stability, as seething unrest ferments ever more widely. In response, many are already striving to devise and implement something better. Even official documents, for instance in the European Union, now allude to the possibility of fostering a 'circular economy', minimizing waste and reusing resources, with business transactions and relationships to match - a significant departure from the prevailing Fire Economy.

But we can go much farther. Why not be truly visionary? The most striking physical difference between the Fire Economy and a future Electric Economy might be to eliminate reliance on violent temperature differences. Fire produces heat so hot it is dangerous. Constructive natural processes, however, especially biological processes, almost all take place within a temperature range that we humans can handle. As yet, we don't know how nature does it. Over sufficient time, nevertheless, we can imagine human-activity systems converging toward natural systems, functioning not with the brute force of fire but with the elegance of electricity. Our global economy and our global society, interacting and interdependent, would be utterly transformed.

Neanderthals had to cherish fire. We have to leave it behind. Infrastructure changes only slowly. But minds can change in an instant. Today could be the day you start thinking beyond the Fire Age.

Afterword

Writing a book is an adventure. You set out confidently, thinking that you know more or less where you are going. But as you progress, the landscape changes. New vistas open, inviting you in unexpected directions. Before long you have almost forgotten where you started, or what path you intended to follow. That happened to me with my previous book, *Keeping The Lights On: Towards Sustainable Electricity*, and it's now happened again. As before, this is not the book I thought I was going to write. It began in 2007, as *Managing Energy: For Climate and Security*. Before long I found that I was 'Rethinking The Fundamentals', and *Rethinking Energy*. For a brief interlude the title became *Keeping Warm and Doing Work: What We Want From Energy and How To Get It*. Then, as further realizations dawned and my focus shifted, it became *The Trouble With Fire*. For another brief interlude it became *Out Of The Stone Age: From Fire To Electricity*, as an unexpected and novel narrative began to take shape in my mind.

With the title *Beyond The Fire Age: What We Do, How We Do It, How We Can Do Better* I at last nearly completed what was for me at least a coherent text with some surprising details. When the draft had reached 25 000 words, not one of the words was either 'energy' or 'climate'. 'Energy' now appears, but very briefly, 'climate' not at all. It was a conscious decision, part of my attempt to avoid tired and misleading language, and to develop a better 'story' about how we got here, where we are and where we might go. Then, as I completed the book, one further change became inevitable. The central theme of the story became vividly clear. It is now embodied in the final title - *Electricity Vs Fire: The Fight For Our Future*.

Because this fight is so urgent, and time is short, I decided to bypass the traditional publishing process, with its timescale in months if not years, and publish the book myself, online. That also allowed me to set a price low enough to make the book easily available to anyone interested. In any case, the informal style and personal content of the book make it unsuitable for publication by my colleagues at Chatham House in London, where I still have the honour to be an Associate Fellow in the Energy, Environment and Resources Programme. I am therefore all the more grateful for their help and support during its long gestation, in the course of which the Programme published several interim working papers for me as I wrestled with the analysis and its implications. Thanks in particular to Programme Heads Bernice Lee and Rob Bailey, and to Gemma Green, Jens Hein, Anna Stapleton, Antony Froggatt, Kirsty Hamilton and Glada Lahn. Thanks also to friends and colleagues around the world who read the first draft and provided a wealth of useful and valuable comment, including Lynn Anderson, Mahesh Bhave, Godfrey Boyle, Joe Court, Lori Davies, Sam Davis, Steve Fawkes, David Fisk, Paul Hofseth, Eric Martinot, Matthew Rhodes, Bent Sorensen, Ian Temperton, Tom Turner, Joanne Wade and Becky Willis. Needless to say, however, any remaining errors and misjudgements are mine alone. My special thanks, too, to Sam Davis for his striking cover design.

My beloved Cleone made my work possible and my life a joy. This is in her memory, and for our daughters Perd and Tab, with all my love.

Walt Patterson
Chesham Bois, Bucks
March 2015

About the author

Walt Patterson is Associate Fellow in the Energy, Environment and Resources Programme at Chatham House in London, UK, and a Visiting Fellow at the University of Sussex. Born in Canada, he has lived in the UK since 1960. A postgraduate nuclear physicist, he has been actively involved in energy and environment since the late 1960s, teaching, writing and campaigning.

His previous book, *Keeping The Lights On: Towards Sustainable Electricity*, was his thirteenth. He has also published hundreds of papers, articles and reviews, on topics including nuclear power, coal technology, renewable energy, energy systems, energy policy and electricity. He has been specialist advisor to two Select Committees of the House of Commons, an expert witness at many official hearings, a frequent broadcaster and advisor to media, and speaker or chair in conferences around the world. He has been awarded the Melchett Medal of the Energy Institute. The *Scientific American 50* named him 'energy policy leader' for his advocacy of decentralized electricity. He is a founding member of the International Energy Advisory Council.

Walt Patterson On Energy, <www.waltpatterson.org>, is an online archive of his work since 1970. It averages over 600 hits a day, with visits from more than 120 countries.

Made in the USA
Charleston, SC
19 May 2015

archive of his work since 1970.

"This is a wonderful, provocative and enlightening book. After more than 50 years working in the 'energy economy', Walt Patterson has come up with a compelling new story. Sweeping back over the whole history of humankind's use of energy, he reveals our day-to-day lives through a different lens, changing the language to help us focus on the differences between fire-fed forms of energy (from coal, oil, gas and even nuclear) and fire-free electricity from the sun and other renewables. In urging us to stop 'feeding the fire' he maps out what is in essence 'the most fundamental transformation' of the way we live on Earth. A revolution, in effect, as we move from the Fire Economy to the Electric Economy."

- Jonathon Porritt, co-founder of Forum For The Future, UK

You can order *Electricity Vs Fire: The Fight For Our Future* as a Kindle e-book, ISBN 978-0-9932612-1-3, from Amazon - £3.00 from Amazon UK, http://www.amazon.co.uk/dp/B00W5HO1RY , $4.40 from Amazon US http://www.amazon.com/dp/B00W5HO1RY , and equivalent prices from Amazon elsewhere with the same page code.

The PDF, ISBN 978-0-9932612-2-0, from Gumroad in the US, https://gumroad.com/l/gKgEJ , is $5.00. They all download directly onto your Kindle, tablet, smartphone or computer.

For readers who prefer a traditional bound book, a print-on-demand edition, ISBN 978-0-9932612-0-6, is now in preparation.

For details, previews and readers' comments see Walt Patterson On Energy, <www.waltpatterson.org>.

ELECTRICITY VS FIRE:
THE FIGHT FOR OUR FUTURE

WALT PATTERSON

Why can't you breathe in Beijing? Why are governments wrangling over the Arctic seabed? Why have we ever more extreme weather worldwide? The answer is *fire*. Fire in heaters, furnaces, engines and power stations is poisoning air in megacities everywhere. Feeding fire is why governments worry about fuel supplies. Fire produces the carbon dioxide upsetting planetary systems. Fire threatens our future.

We think of fire as welcoming. But fire is violent, extreme, rapidly turning resources into wastes, toxic and pernicious. Yet we still rely on fire, even when we need not, and despite the ever intensifying problems fire creates. To address pollution, security, and climate we need to minimize human use of fire.

Fire, however, has let us control electricity. Electricity, in turn, may save us from fire. *Electricity Vs Fire is The Fight For Our Future.*

Walt Patterson is Associate Fellow in the Energy, Environment and Resources Programme at Chatham House in London, UK, a Visiting Fellow at the University of Sussex and a founder-member of the International Energy Advisory Council. *Electricity Vs Fire: The Fight For Our Future* is his fourteenth book.